PRAIRIE PLANTS
AND THEIR ENVIRONMENT

Prairie Plants
and their Environment

A Fifty-Year Study in the Midwest

by

J. E. Weaver

UNIVERSITY OF NEBRASKA PRESS · LINCOLN

Copyright © 1968 by the University of Nebraska Press

All rights reserved

Library of Congress Catalog Card Number 67–19160

Publisher's Preface

From 1916 until his death in 1966, John E. Weaver studied the grasslands of the central United States and the ecology of their component species with a single-mindedness of purpose few biologists can match. Two weeks before his death at the age of eighty-two he submitted to the University of Nebraska Press the manuscript of a book which was to be, he said, the final statement drawn from a lifetime of ecological research. The present book is that statement.

The original manuscript has not been altered except to change minor grammatical errors. The structure and content of the book are Professor Weaver's. If the manuscript described work finished in 1934 as having been done "recently," the adverb was retained, although much of ecological importance has occurred in the intervening thirty years to modify the conclusions. In cases where data are cited without notation, it only can be assumed that they are Professor Weaver's or one of his associate's and that the original experiments may be found in one of his more than one hundred scientific reports.

On page 48 he summarizes the purposes of his investigations: "to clarify some of the many problems presented by this vast natural unit of vegetation, to better understand the importance and significance of grassland and its utilization, and to furnish a permanent record of a rapidly vanishing vegetation." His research did, indeed, meet his objectives. Throughout the pages of this book the skill and zeal of this pioneer plant ecologist are revealed. His entire life was devoted to, as he himself says, "a most interesting and fascinating task."

Forty-two students completed Ph.D. degrees under his direction and about fifty students did their master's thesis work with him in the Department of Botany at the University of Nebraska.

The following bibliography of his published work was compiled by Professor Weaver some months before his death and is believed to be a complete listing of his works, exclusive of reviews, abstracts, year-book reports, and the like.

Bibliography of John E. Weaver

Weaver, J. E. 1914. Evaporation and plant succession in southeastern Washington and adjacent Idaho. Plant World, 17:273–294.

Humphrey, H. B., and J. E. Weaver. 1915. Natural reforestation in the mountains of northern Idaho. Plant World, 18:31–47.

Weaver, J. E. 1915. A study of the root-systems of prairie plants of southeastern Washington. Plant World, 18:227–248; 273–292.

———. 1916. The effects of certain rusts upon the transpiration of their hosts. Minn. Bot. Studies, 4:379–408.

———. 1917. A study of the vegetation of southeastern Washington and adjacent Idaho. Univ. [Nebr.] Studies, 17:1–133.

———, and A. F. Thiel. 1917. Ecological studies in the tension zone between prairie and woodland. Bot. Surv. Nebr. n.s., 1:1–60.

Pool, R. J., J. E. Weaver, and F. C. Jean. Further studies in the ecotone between prairie and woodland. Univ. [Nebr.] Studies, 18:1–47.

Weaver, J. E. 1918. The quadrat method in teaching ecology. Plant World, 21:267–283.

———. 1919. The ecological relations of roots. Carnegie Inst. Wash. Pub. No. 286, 128 p.

———, and A. Mogensen. 1919. Relative transpiration of coniferous and broad-leaved trees in autumn and winter. Bot. Gaz. 68:393–424.

———. 1920. Root development in the grassland formation. Carnegie Inst. Wash. Pub. No. 292, 151 p.

———, F. C. Jean, and J. W. Crist. 1922. Development and activities of roots of crop plants. Carnegie Inst. Wash. Pub. No. 316, 117 p.

———, and J. W. Crist. 1922. Relation of hardpan to root penetration in the Great Plains. Ecology, 3:237–249.

Jean, F. C., and J. E. Weaver. 1924. Root behavior and crop yield under irrigation. Carnegie Inst. Wash. Pub. No. 347, 65 p.

Clements, F. E., and J. E. Weaver. 1924. Experimental vegetation. Carnegie Inst. Wash. Pub. No. 355, 172 p.

Weaver, J. E., J. Kramer, and M. Reed. 1924. Development of root and shoot of winter wheat under field environment. Ecology, 5:26–50.

Bruner, W. E., and J. E. Weaver. 1924. Size and structure of leaves of cereals in relation to climate. Univ. [Nebr.] Studies, 24:1–37.

Crist, J. W., and J. E. Weaver. 1924. Absorption of nutrients from subsoil in relation to crop yield. Bot. Gaz., 77:121–148.

Weaver, J. E. 1924. Plant production as a measure of environment: a study in crop ecology. J. Ecology, 12:205–237.

———, and J. W. Crist. 1924. Direct measurement of water loss from vegetation without disturbing the normal structure of the soil. Ecology, 5:153–170.

———, H. C. Hanson, and J. M. Aikman. 1925. Transect method of studying woodland vegetation along streams. Bot. Gaz., 80:168–187.

———. 1925. Investigations on the root habits of plants. Amer. Jour. Bot., 12:502–509.

———. 1926. Root development of field crops. McGraw-Hill Book Co. Inc., New York, 291 p.

———, and W. E. Bruner. 1927. Root development of vegetable crops. McGraw-Hill Book Co., Inc., New York. 351 p.

Bibliography of John E. Weaver   vii

————. 1927. Some ecological aspects of agriculture in the prairie. Ecology, 8:1–17.

————, and F. E. Clements. 1929. Plant ecology. McGraw-Hill Book Co., Inc., New York. 520 p.

Clements, F. E., J. E. Weaver, and H. C. Hanson. 1929. Plant competition. Carnegie Inst. Wash. Pub. No. 398, 340 p.

Weaver, J. E., and J. W. Himmel. 1929. Relation between the development of root system and shoot under long- and short-day illumination. Plant Physiol., 4:435–457.

————. 1930. Underground plant development in relation to grazing. Ecology, 11:543–557.

————, and W. J. Himmel. 1930. Relation of increased water content and decreased aeration to root development in hydrophytes. Plant Physiol., 5:69–92.

————. 1931. Who's who among prairie grasses. Ecology, 12:623–632.

————, and W. J. Himmel. 1931. The environment of the prairie. Univ. Nebr. Cons. and Surv. Div. Bull., 5:1–50.

————, and T. J. Fitzpatrick. 1932. Ecology and relative importance of the dominants of tall-grass prairie. Bot. Gaz. 93:113–150.

————, and J. Kramer. 1932. Root system of *Quercus macrocarpa* in relation to the invasion of prairie. Bot. Gaz. 96:51–85.

Biswell, H. H., and J. E. Weaver. 1933. Effect of frequent clipping on the development of roots and tops of grasses in prairie sod. Ecology, 14:368–390.

Weaver, J. E., and T. J. Fitzpatrick. 1934. The prairie. Ecol. Monog., 4:109–295.

————, and E. L. Flory. 1934. Stability of climax prairie and some environmental changes resulting from breaking. Ecology, 15:333–347.

————, and G. W. Harmon. 1935. Quantity of living plant materials in prairie soils in relation to runoff and soil erosion. Univ. Nebr. Cons. and Surv. Div. Bull. 8, 53 p.

————, V. H. Hougen, and M. D. Weldon. 1935. Relation of root distribution to organic matter in prairie soils. Bot. Gaz. 96:389–420.

————, and W. C. Noll. 1935. Comparison of runoff and erosion in prairie, pasture, and cultivated land. Univ. Nebr. Cons. and Surv. Div. Bull. 11, 37 p.

————. 1935. Measurement of runoff and soil erosion by a single investigator. Ecology, 16:1–12.

————, L. A. Stoddart, and W. Noll. 1935. Response of the prairie to the great drought of 1934. Ecology, 16:612–629.

————, and J. Kramer. 1935. Relative efficiency of roots and tops of plants in protecting the soil from erosion. Science, 82:354–355.

————, and F. W. Albertson. 1936. Effects of the great drought on the prairies of Iowa, Nebraska, and Kansas. Ecology, 17:567–639.

Kramer, J., and J. E. Weaver. 1936. Relative efficiency of roots and tops of plants in protecting the soil from erosion. Univ. Nebr. Cons. and Surv. Div. Bull. 12, 94 p.

Weaver, J. E., and F. E. Clements. 1938. Plant Ecology, 2nd edition. McGraw-Hill Book Co., Inc., New York. 601 p.

————, and V. H. Hougen. 1939. Effect of frequent clipping on plant production in prairie and pasture. Amer. Midl. Nat., 21:396–414.

————, and F. W. Albertson. 1939. Major changes in grassland as a result of continued drought. Bot. Gaz., 100:576–591.

Bukey, F. S., and J. E. Weaver. 1939. Effects of frequent clipping on the underground food reserves of certain prairie grasses. Ecology, 20:246–252.

Shively, S. B., and J. E. Weaver. 1939. Amount of underground plant materials in different grassland climates. Univ. Nebr. Cons. and Surv. Div. Bull. 21, 68 p.

Weaver, J. E., and W. W. Hansen. 1939. Increase of *Sporobolus cryptandrus* in pastures of eastern Nebraska. Ecology, 20:374–381.

————, and F. W. Albertson. 1940. Deterioration of grassland from stability to denudation with decrease in soil moisture. Bot. Gaz., 101:598–624.

————. 1940. Deterioration of midwestern ranges. Ecology, 21:216–236.

Fowler, R. L., and J. E. Weaver. 1940. Occurrence of a disease of sideoats grama. Bull. Torrey Bot. Club, 67:503–508.

Weaver, J. E., J. H. Robertson, and R. L. Fowler. 1940. Changes in true prairie during drought as determined by list quadrats. Ecology, 21:357–362.

————, and W. W. Hansen. 1941. Native midwestern pastures: their origin, composition, and degeneration. Univ. Nebr. Cons. and Surv. Div. Bull. 22, 93 p.

————. 1941. Regeneration of native midwestern pastures under protection. Univ. Nebr. Cons. and Surv. Div. Bull. 23, 91 p.

Albertson, F. W., and J. E. Weaver. 1942. History of the native vegetation of western Kansas during seven years of continuous drought. Ecol. Monog., 12:23–51.

Weaver, J. E. 1942. Competition of western wheat grass with relict vegetation of prairies. Amer. J. Bot., 29:366–372.

————, and I. M. Mueller. 1942. Role of seedlings in recovery of midwestern ranges from drought. Ecology, 23:275–294.

Mueller, I. M., and J. E. Weaver. 1942. Relative drought resistance of seedlings of dominant prairie grasses. Ecology, 23:387–398.

Weaver, J. E., and F. W. Albertson. 1943. Resurvey of grasses, forbs, and underground plant parts at the end of the great drought. Ecol. Monog., 13:64–117.

————. 1943. Replacement of true prairie by mixed prairie in eastern Nebraska and Kansas. Ecology, 24:421–434.

————. 1944. Recovery of midwestern prairies from drought. Proc. Amer. Phil. Soc., 88:125–131.

————, and R. W. Darland. 1944. Grassland patterns in 1940. Ecology, 25:202–215.

Albertson, F. W., and J. E. Weaver. 1944. Effects of drought, dust, and intensity of grazing on cover and yield of short-grass pastures. Ecol. Monog., 14:1–29.

Weaver, J. E., and F. W. Albertson. 1944. Nature and degree of recovery of grassland from the great drought of 1933 to 1940. Ecol. Monog., 14:393–479.

————. 1944. North American prairie. Amer. Scholar, 13:329–339.

————, and R. W. Darland. 1945. Yields and consumption of forage in three prairie types: an ecological analysis. Nebr. Cons. Bull. 27, 76 p.

Albertson, F. W., and J. E. Weaver. 1945. Injury and death or recovery of trees in prairie climate. Ecol. Monog., 15:393–433.

Weaver, J. E., and W. E. Bruner. 1945. A seven-year quantitative study of succession in grassland. Ecol. Monog., 15:297–319.

————, and E. Zink. 1945. Extent and longevity of the seminal roots of certain grasses. Plant Physiol. 20:359–379.

————. 1946. Annual increase of underground materials in range grasses. Ecology, 27:115–127.

————. 1946. Length of life of roots of ten species of perennial range and pasture grasses. Plant Physiol., 21:201–217.

Albertson, F. W., and J. E. Weaver. 1946. Reduction of ungrazed mixed prairie to short grass as a result of drought and dust. Ecol. Monog., 16:449–463.

Weaver, J. E. 1947. Rate of decomposition of roots and rhizomes of certain range grasses in undisturbed prairie soil. Ecology, 28:221–240.

————, and R. W. Darland. 1947. A method of measuring vigor of range grasses. Ecology, 28:146–162.

————. 1948. Changes in vegetation and production of forage resulting from grazing lowland prairie. Ecology, 29:1–29.

————, and W. E. Bruner. 1948. Prairies and pastures of the dissected loess plains of central Nebraska. Ecol. Monog., 18:507–549.

————, and R. W. Darland. 1949. Quantitative study of root systems in different soil types. Science, 110:164–165.

————. 1949. Soil-root relationships of certain native grasses in various soil types. Ecol. Monog., 19:303–338.

————, and J. Voigt. 1950. Monolith method of root-sampling in studies on succession and degeneration. Bot. Gaz. 111:286–299.

————. 1950. Effects of different intensities of grazing on depth and quantity of roots of grasses. J. Range Mgt., 3:100–113.

————. 1950. Stabilization of midwestern grasslands. Ecol. Monog., 20:251–270.

Voigt, J. W., and J. E. Weaver. 1951. Range condition classes of native midwestern pasture; an ecological analysis. Ecol. Monog., 21:39–60.

Weaver, J. E., and G. W. Tomanek. 1951. Ecological studies in a midwestern range: the vegetation and effects of cattle on its composition and distribution. Univ. Neb. Cons. and Surv. Div. Bull. 11, 82 p.

————, and N. W. Rowland. 1952. Effects of excessive mulch on development, yield, and structure of native grassland. Bot. Gaz., 114:1–19.

Branson, F. W., and J. E. Weaver. 1953. Quantitative study of degeneration of mixed prairie. Bot. Gaz., 114:397–416.

Fox, R. L., J. E. Weaver, and R. C. Lipps. 1953. Influence of certain profile characteristics upon the distribution of the roots of grasses. Agron. J., 45:583–589.

Weaver, J. E. 1954. A seventeen-year study of plant succession in prairie. Amer. J. Bot. 41:31–38.

————, and W. E. Bruner. 1954. Nature and place of transition from true prairie to mixed prairie. Ecology, 35:117–126.

————. 1954. North American Prairie. Johnsen Pub. Co., Lincoln, Nebr., 348 p.

————, and F. W. Albertson. 1956. Grasslands of the Great Plains, their nature and use. Johnsen Pub. Co., Lincoln, Nebr., 395 p.

————. 1958. Summary and interpretation of underground development in natural grassland communities. Ecol. Monog., 28:55–78.

————. 1958. Classification of root systems of forbs of grassland and a consideration of their significance. Ecology, 39:393–401.

————. 1958. Native grasslands of southwestern Iowa. Ecology, 39:733–750.

————. 1960. Floodplain vegetation of the central Missouri Valley and contacts of woodland with prairie. Ecol. Monog., 30:37–64.

————. 1960. Comparison of vegetation of the Kansas-Nebraska drift loess hills and loess plains. Ecology, 41:73–88.

————. 1960. Extent of communities and abundance of the most common grasses in prairie. Bot. Gaz., 122:25–33.

————. 1961. Return of midwestern grassland to its former composition and stabilization. Occas. Papers Adams Ctr. Ecol. Studies, 3:1–15.

————. 1961. The living network in prairie soils. Bot. Gaz., 123:16–28.

————. 1963. The wonderful prairie sod. J. Range Mgt., 16:165–171.

————. 1965. Native vegetation of Nebraska, University of Nebraska Press, Lincoln, 185 p.

Contents

PRAIRIE PLANTS
AND THEIR ENVIRONMENT

I.

The Good Earth and What It Contains

When, in 1916, the writer came to the University of Nebraska as a young instructor, he came with a background that well fitted him for studies in the field. His research in the grassland and adjoining forests of southeastern Washington and adjacent Idaho had been completed and a fine acquaintance had been made with methods of examining the prairie both above and below ground level (Weaver, 1915, 1917). Similar studies in the area between the Missouri River and the Rocky Mountains were now to be made, and these have resulted in many new investigations that are of both scientific and practical value.

During 1914 the writer had investigated the root systems of the prairie plants of southeastern Washington, where an annual precipitation of only 21.6 inches occurred, mostly in the period of rest. After that he planned to make a comparative study of the roots of prairie plants in a more humid region, where the precipitation occurs mostly during the season of plant growth. The opportunity for such a study came during the fall of 1917, and the first publication of results appeared in "The Ecological Relations of Roots," in 1919. Here may be found descriptions of the character, depth, and distribution of the roots of about 140 species of grasses, shrubs, and forbs of prairie and plains. A companion work, "Root Development in the Grassland Formation"—also published by the Carnegie Institution of Washington —appeared in 1920. It includes the results of investigations at more than 25 stations in Kansas, Colorado, South Dakota, and Nebraska.

The climax plant community has integrated all of the environmental factors of its habitat; it is the fundamental response to the controlling conditions. The individual root habit and especially the community root habit, together with the more familiar above-ground parts, serve to interpret the environmental conditions. Both of these criteria are needed to reveal the "judgment" of the plants as to the fitness of the habitat in which they grow or in which crop plants are to be grown.

Frequently, half—and often much more—of every plant in grassland is invisible. This is the part in the soil (see Fig. 1). It is the most

1

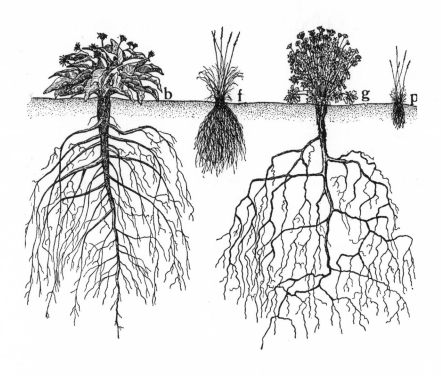

Fig. 1.—Root systems of plants from the Palouse prairie of southeastern Washington. The picture emphasizes root depths, as deep as 6 feet, compared with tops, which are about 2 feet high. *b, Balsamorhiza sagittata; f, Festuca ovina ingrata; g, Geranium viscosissimum; p, Poa sandbergii.*

permanent part and changes but little from season to season. Conversely, the parts above ground—the image we recall as prairie—cease growth and die with the approach of winter, and for half the year the living prairie is underground. The soil is a part of the environment in which plants grow and upon which plant life is very dependent. Soil is usually defined as the unconsolidated outer layer of the earth's crust which, through processes of weathering and the incorporation of organic matter, becomes adapted to plants. In grassland, the soil usually contains and acts upon a more extensive part of the plant body than does the atmosphere. The true soil or solum is usually composed of the parent materials which it covers. The introduction of living material is largely responsible for the constructional processes of soil development since residues of plants and animals return to the soil more than green plants take away. As a result, the soil contains stored-up energy and becomes the abode of bacteria, fungi, actinomycetes, and many other microorganisms.

Soil is not nearly as compact as it seems. An upland soil near Lincoln, Nebraska, covered with little bluestem, was composed of 43 to 50 percent solid matter in the surface foot; the remainder was pore space. The pore space was about equally divided between air space and space that was occupied by average water content during the growing season. Even at a depth of 4 feet, the solid matter was only 60 percent and the air space was 20 percent. Under big bluestem on a well-drained flood plain, solid matter varied from 40 to 50 percent from the surface to 7 feet in depth, but the air space was reduced with depth from 28 to 10 percent. Many grassland plants extend their roots to even greater depths.

SOILS OF PRAIRIE AND PLAINS

The major soil groups in which field studies have been made are Brunizem, Chernozem, Chestnut, and Brown soils. These are all soils that have developed under a cover of grass.

Brunizems occupy all of western Iowa, but in Nebraska they are limited to the eastern part of the state (see Fig. 2). These soils are usually about 2.5 to 3 feet deep, and beneath them the parent materials of loess or glacial drift (from which Brunizems were formed) extend many feet deeper. Both soil and parent materials are usually moist.

The upper or A horizon of Brunizems of eastern Nebraska and western Iowa is usually about 18 inches thick. This topsoil has a granular structure which results from relatively high precipitation (in excess of 25 inches), repeated freezing and thawing, and alternate wetting and drying. The high humus content (2 to 6 percent) and the favorable effects of root activities also are important features which combined to produce the granular structure. Roots penetrate the topsoil easily.

The subsoil or B horizon usually extends to a depth of about 30 inches. Root penetration is much more difficult here and branching is less pronounced because of a higher clay content and prismatic structure. The C horizon of parent materials, from which the soil develops, has a lower clay content than the subsoil. It is moderately mellow and roots penetrate much more easily, often to depths of several feet.

Soil scientists have summarized the major soil-forming processes in Brunizems as the accumulation of organic matter in the surface layer, the leaching of bases (lime, etc.) and development of acidity in the A and B horizons, the formation of a high cation-exchange capacity clay, and the accumulation of this clay in the B horizon (Riecken, 1960).

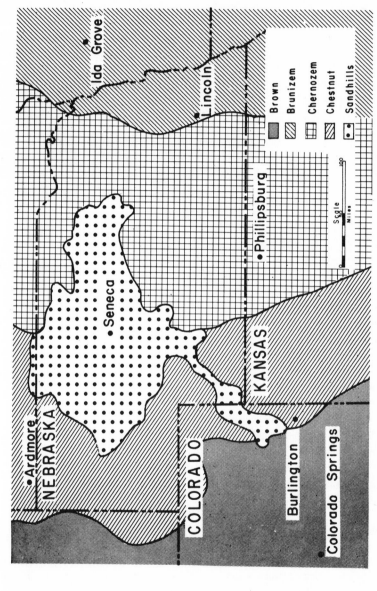

Fig. 2.—Major soil groups in the area of these studies and the chief sandhill area. Conservation and Survey Division, University of Nebraska.

The relation of root distribution to organic matter in prairie soils has been studied by Weaver, Hougen, and Weldon (1935).

Chernozem soils develop under a continental climate of less than 25 inches of rainfall and, usually, an excess of evaporation (as measured from a free water surface) over precipitation. Their eastward part is clothed with true prairie whereas, under lighter rainfall, mixed prairie prevails westward. Thus Chernozems are found in the more humid part of the dry region, as at Phillipsburg, Kansas (see Fig. 2). The luxuriant growth of mid and short grasses has produced black soil that is very high in its content of organic matter and very fertile for cultivated cereal crops, such as wheat and sorghums.

The soil-forming processes that lead to the development of Chernozems are mainly the addition of organic matter to the A soil horizon, the formation of clay and development of structure in the B horizon or subsoil, and the movement of bases (lime, etc.) by leaching to a zone of accumulation in the lower part of the solum (Elder, 1960). Although the soil below the lime layer is often dry, sometimes the earth is moist to many feet in depth. These fine-textured soils are composed of sands (mostly fine sand), silt, and clay. They are fertile and productive because they contain appreciable quantities of easily weathered minerals which represent reserves of mineral plant nutrients. The clays favor retention of available and exchangeable plant nutrients. The climate favors vigorous root action and microbial activity, which are important agencies in mineral weathering. High productivity is due to a combination of soil and favorable climate.

West of the Chernozem soil belt are found the Chestnut soils of western Nebraska and the Brown soils of eastern Colorado. They occur in an area of 21 inches or less rainfall, as at Burlington, Colorado (see Fig. 2). With decreased precipitation the vegetation becomes sparser, the soil becomes lighter in color, and the solum becomes thinner. In some places the Brown soil is only 6 to 12 inches deep. The vegetation of this soil, while not luxuriant, is still well developed, especially the parts underground. The deeper layer of soil, with less humus, is lighter brown and contains the calcium accumulations. The lime zone decreases in depth with decreased precipitation, and the influence of the vegetation on the soil farther westward becomes less and less. The fertile Chestnut soils and some of the Brown soils are used for wheat production, despite the hazardous, climate. Since the nutrients have not been leached, the soils, though low in organic matter, are fertile. Available water content, even of the upper soil, is often low; and usually, but not always, parent materials below the soil are dry.

UNDERGROUND PLANT PARTS IN TRUE PRAIRIE

The grasslands east of 98° 30′ west longitude receive more precipitation than those of the Great Plains west of this line, and the former are composed mostly of tall grasses 5 to 8 feet in height and mid grasses 2 to 4 feet tall. These, in sequence, are illustrated by big bluestem (*Andropogon gerardi*) and little bluestem (*A. scoparius*). This portion of the grassland is usually designated as true prairie, in contrast to the mixed (mid- and short-grass) prairie in the less favorable climate westward. Short grasses are only 0.5 to 1.5 feet high.

In the study of underground parts the following method was employed. A trench 2.5 feet in width and 5 to 10 feet long was dug to a

FIG. 3.—One end of the first trench used for the study of root systems. Pullman, Washington, 1914.

FIG. 4.—From left to right are Indian grass (*Sorghastrum nutans*), little bluestem (*Andropogon scoparius*), and false boneset (*Kuhnia eupaortioides*).

depth of about 5 feet by the side of the plants to be examined. This afforded an open face into which one might dig with a hand pick (furnished also with a cutting edge) and an ice pick to make an examination of the root systems. The trenches, of course, were deepened as necessary, sometimes to 15 or more feet (Fig. 3). More than 150 such excavations were made during the four years of this study, in which more than 1,500 root systems were examined. This apparently simple process of root excavation, however, requires much practice, not a little patience, and wide experience with soil texture.

The root descriptions, except as otherwise indicated, are all of mature plants. Many root systems, especially those of grasses, were removed and photographed; others were drawn. In the drawings the root systems were shown as nearly as possible in a vertical plane. Drawings were made simultaneously with the excavation of the roots and always to exact measurements. Indeed, such drawings, carefully executed, often represent the extent, position, and minute branching more accurately than even a photograph.

The roots and tops of three important plants are shown in Figure 4. Indian grass (*Sorghastrum nutans* [left]) is about 6 feet deep and little bluestem (*Andropogon scoparius* [center]) is about 4 feet deep. The root system of false boneset (*Kuhnia eupatorioides*) is shown on the right; it is an upland forb which often attains a depth of 16 feet. The first 5 feet of the roots of big bluestem (*Andropogon gerardi*) and switchgrass (*Panicum virgatum*) are shown in Figure 5. These roots are from monoliths of soil 12 inches wide and 5 feet long.

The taller and deeper-but-coarser-rooted *Andropogon gerardi* requires more moist soil than the shorter, more finely rooted *A. scoparius*; this is shown both by their local and general distribution. The former luxuriates in draws and on lower lands while the latter dominates drier areas. Westward, the big bluestem is the first to disappear.

Twelve plants of big bluestem were examined near Lincoln, Nebraska. On lowlands, the very abundant roots descend almost vertically from the base of the plant and its network of rhizomes, and at once thoroughly occupy the soil and form a dense sod. But on uplands, in drier soil, some roots may extend obliquely from the bunches for more than a foot before turning downward. The larger roots vary from 0.5 to 3 millimeters in diameter and may reach a depth of 7 feet. All of the roots branch profusely; the main laterals are from 2 to 6 inches long. The ends of these reddish-brown roots are extremely well branched to the tip.

Roots of *Andropogon scoparius* are much finer than those of the preceding species and are only 0.1 to 0.8 millimeters in diameter. A lateral spread of 1.5 feet in the surface foot of soil is usual and a depth

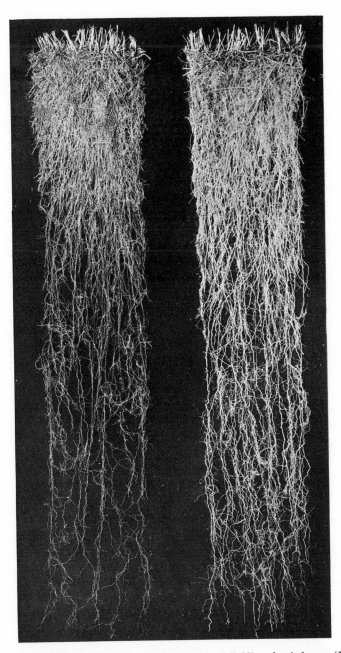

Fig. 5.—Roots of big bluestem (*Andropogon gerardi* [left]) and switchgrass (*Panicum virgatum*) from monoliths of soil 12 inches wide, 3 inches thick (into the trench wall), and 5 feet deep. The bluestem was 7 feet and the switchgrass more than 8 feet deep. From Weaver and Darland, *Ecological Monographs*, 1949a.

of 5 feet is ordinarily attained. The upper 2 to 3 feet of soil is especially well occupied, but branching is profuse almost to the root tips. All of the roots branched profusely to the third or fourth order, and many of the branches were more than 30 inches long. The roots are light brown in color and have a very thick cortex which peels off easily, thus exposing the tough, yellowish stele.

Sorghastrum nutans, Indian grass, occurs—usually in small amounts —both on uplands and lowlands. It is one of the deeply rooted prairie grasses. The roots are intermediate in coarseness and depth, as compared to those of the two bluestems.

Prairie cordgrass (*Spartina pectinata*) is a very tall, coarse plant of wet soils. Its rank growth, to a height of 8 or more feet, and dense sod-forming habit frequently excludes almost all other vegetation. It is an indicator of land with a high water content and poor soil aeration. Roots descend from the base of the stems and from the very extensive network of coarse rhizomes. They are very coarse and tough, and some measure 3 or 4 millimeters in diameter. They taper so gradually that, at a depth of 7 to 8 feet, they are 1 to 1.5 millimeters thick. Some roots were traced to a depth of 9 feet, and may have extended a foot or two deeper. Branches may run off horizontally for 5 to 6 inches and then turn downward. Laterals are threadlike and very abundant, but usually are short, crooked, and poorly rebranched. This grass has the coarsest roots of all plants examined. Its great depth of penetration is surprising, until one recalls that swamp lands frequently dry out in late summer and produce great cracks in the soil, especially in dry years. An enormous top growth demands much water.

Switchgrass (*Panicum virgatum*) usually occupies soil with a water content that is intermediate between that of big bluestem on the drier side and prairie cordgrass on the wetter side. In fact, it often forms extensive communities in just such areas. It is a tall, coarse, sod-forming species, and has the longest root system of all the grasses examined. Roots arise from stem bases and from extensive rhizomes and penetrate more or less vertically downward to depths of 9 to 11 feet. Branching of the rather coarse roots is not profuse, nor are the branches extensive. The extent of foliage and roots (to a depth of 9 feet) is shown in Figure 6. When flower stalks are fully developed and flowering occurs, heights of 6 to 10 feet are attained.

Upland grasses, typically bunch grasses, usually spread widely in the upper soil. The roots intermingle with those of adjacent bunches. Junegrass (*Koeleria cristata*) is a rather small, short-lived, perennial with roots to a depth of 1.5 to 2 feet. Aside from little bluestem, needlegrass (*Stipa spartea*) and prairie dropseed (*Sporobolus heterolepis*) are the most

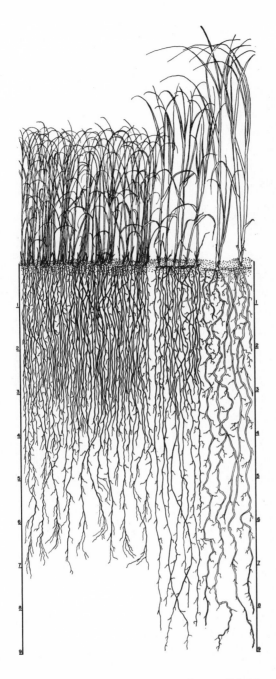

Fig. 6.—Characteristic development of tops and roots of big bluestem (*Andropogon gerardi* [left]), switchgrass (*Panicum virgatum* [center right]), and prairie cordgrass (*Spartina pectinata* [right]). When flower stalks are fully developed and flowering occurs, heights of 6 or more feet are attained.

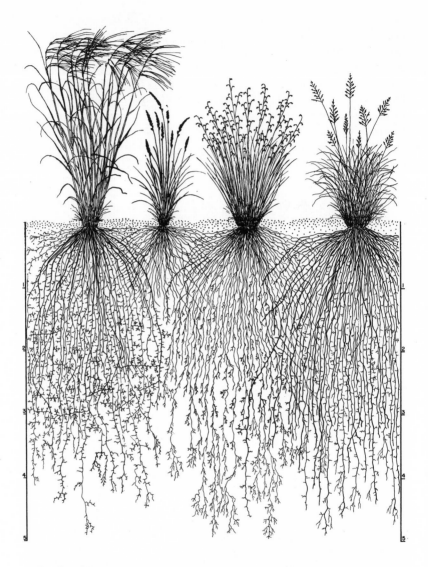

Fig. 7.—Characteristic development of the tops and roots of four bunch grasses as they occur in several upland communities. From left to right they are needlegrass (*Stipa spartea*), Junegrass (*Koeleria cristata*), little bluestem (*Andropogon scoparius*), and prairie dropseed (*Sporobolus heterolepis*). Note that the tops are only about half as high as the roots are deep.

abundant. In Figure 7 (a drawing made in 1960) the relationship of these plants to each other, as regards both roots and tops, is portrayed. The depth of roots is the average for many plants of each species. The roots of all these plants are moderately fine, well branched, and of medium depth—about twice as deep as the plants are tall. The largest roots have only about one-third the diameter of those of big bluestem. Needlegrass and Junegrass are examples of cool-season grasses that flower early, but most upland prairie species are warm-season grasses that flower late, after a long season of growth. In Figures 6 and 7, only the amount of tops that originated in a strip of soil 1 inch in width are shown. Likewise, the average number of roots is from an inch-wide column of soil. Root systems of 15 prairie grasses were examined and described earlier (Weaver, 1919, 1920).

Fig. 8.—Root system of false indigo (*Baptisia bracteata* [left]) and prairie rose (*Rosa suffulta*). Roots of the rose are about 20 feet deep.

A very large number of nongrasslike herbs occur regularly in the prairie. These herbaceous plants are designated as forbs and the root development of many species has been studied and described. The very extensive taproot of a blazing star (*Liatris punctata*) is 16 feet deep. Seven individuals of this species were examined. Ring counts in the woody crown showed that one plant was more than 35 years old. *Liatris scariosa* is characterized by a large, woody corm, 3 to 5 inches in diameter, from which arise numerous, wide-spread, fibrous roots. Another very common composite is the many-flowered aster (*Aster ericoides*). Clumps of this rather bushy plant, 2 to 3 feet high, are connected by tough, woody rhizomes 2 to 4 millimeters thick and 6 to 10 inches long. It is conspicuous in the prairie from the time its white or purplish flowers begin to bloom, in August, until late in October.

The roots of false indigo (*Baptisia bracteata*) and the prairie rose (*Rosa suffulta*) are shown in Figure 8. Wild licorice (*Glycyrrhiza lepidota*) is another deeply rooted forb. This characteristic prairie legume has many-branched rhizomes, which are several feet long, in addition to deep, fleshy roots. The root system of the lead plant (*Amorpha canescens*) is equally extensive. This plant is the most abundant legume in the prairie.

The root systems of many species have been examined in true prairie and many pages were required to display the drawings. A group of 11 representative species is shown in Figure 9. The picture is that of an imaginary excavation in the prairie, 28 feet wide, of similar length, and 20 feet deep. For clarity, the network of grass roots has been omitted and only an inch or two of the topsoil is shown. The root systems are of natural width and length and were redrawn from "Ecological Relations of Roots" and "Root Development in the Grassland Formation." If the picture is inverted, one gets a better idea of the considerable space between the forbs, which, of course, is occupied by grasses. The abundant, finer branches of roots could not be shown because of the greatly reduced size.

The most obvious conclusion from a consideration of the preceding data is that prairie species are provided with well-developed, deep-seated, and extensive root systems. On the basis of root depth, the 43 species examined may be divided into three groups. The first group includes plants with shallow roots that seldom extend below the first 2 feet of soil. This group, consisting entirely of grasses, makes up only 14 percent of the total. The second group, of intermediate depth, is composed of grasses and forbs with roots that extend well below the second foot but seldom deeper than 5 feet—21 percent of prairie

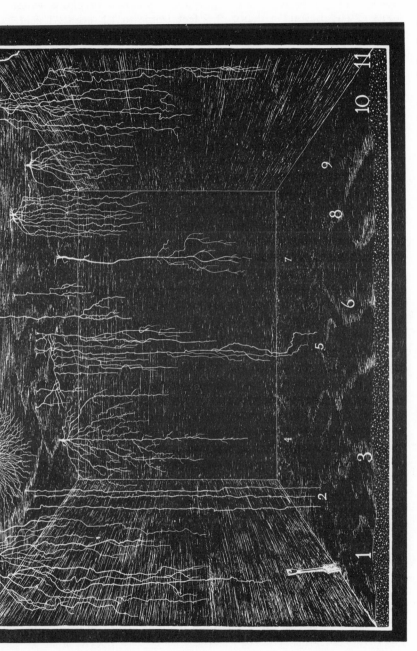

FIG. 9.—Excavation, 20 feet deep, showing the root habits of several prairie forbs. *1, Amorpha canescens*, lead plant; *2, Lygodesmia juncea*, rushlike lygodesmia; *3, Liatris squarrosa*, scaly blazing star; *4, Liatris punctata*, dotted button snakeroot; *5, Rosa suffulta*, prairie rose; *6, Echinacea pallida*, pale-purple coneflower; *7, Silphium laciniatum*, compassplant; *8, Aster ericoides*, many-flowered aster; *9, Astragalus crassicarpus*, ground plum; *10, Glycyrrhiza lepidota*, licorice; *11, Psoralea esculenta*, prairie turnip (turning aside into wall). Redrawn from Weaver (1919–1920), and originally published in the *Botanical Gazette*.

species. The third and largest group is composed of plants whose roots extend beyond a depth of 5 feet (some to 12, even 23, feet), and this group includes 65 percent of the species selected as typical of the prairie flora. Examination of the deeply rooted species shows that only about one-fifth rely—to any marked degree—upon the shallow soil for water or nutrients, and that many species—when mature— carry on relatively little absorption in the first, second, or third foot. Layering of the roots reduces competition and permits the growth of a larger number of species. At an earlier date (Weaver, 1917, 1920), the abundance of rhizomes, corms, bulbs, and root offshoots was given little attention, but their importance will be considered later.

PLANTS OF MIXED PRAIRIE HARDLANDS

Extensive examinations have been made of the short and mid grasses of the mixed prairie. Only species that grow in non-sandy lands, commonly called hardlands, are considered here, and the following section treats the species found in sandy soils.

Buffalo grass (*Buchloe dactyloides*) was excavated at eight widely separated stations in South Dakota, Colorado, Kansas, and Nebraska. Depth of root penetration varied from about 4.5 to 7 feet in Colorado, and was 6 feet at Lincoln, Nebraska. Figure 10 shows the usual wide-spreading roots in the surface soil, the dense mat of roots, and the thorough occupancy of the soil to a depth of 4 to 5 feet. The tough, wiry, main roots are less than 1 millimeter in diameter; the tiny branches usually are only a half-inch in length and poorly branched.

Blue grama (*Bouteloua gracilis*) has a root system very much like that of buffalo grass. Roots spread outward 12 to 18 inches in the surface soil and, like those of buffalo grass, are able to benefit from moderately light showers. Depths of 4 to 6 feet are common. Thus both of these widely spread, sod-forming dominants are short only above the ground. They are only 0.5 to 1.5 feet tall.

Western wheatgrass (*Agropyron smithii*) is a mid grass with a height of 2 to 2.5 feet in the Great Plains. Its stems are connected by tough rhizomes. It has an excellent surface-absorbing system that consists of very numerous and short but extremely well-branched horizontal roots. The main roots are 1.5 to 2 millimeters thick; they are coarse, and have profusely branched laterals.

It is a notable fact that all three of these plains grasses, when grown in true prairie, develop surface roots only poorly, if at all. Maximum root depth varies from 5 to 7 feet.

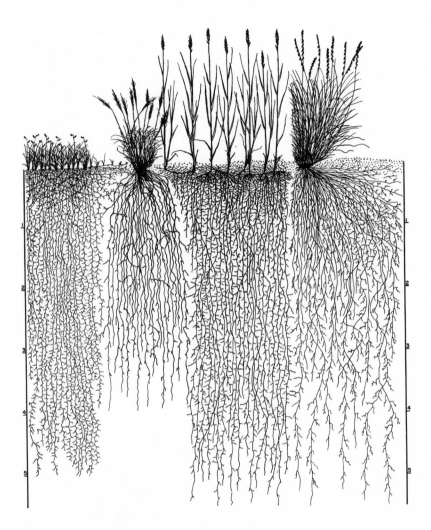

Fig. 10—Tops and roots of dominant grasses common to mixed prairie (only the parts in a strip 1 inch wide are shown). From left to right are buffalo grass (*Buchloe dactyloides* [height and depth of which is similar to that of blue grama]), purple three-awn (*Aristida purpurea*), western wheatgrass (*Agropyron smithii*), and side-oats grama (*Bouteloua curtipendula*).

Drawings of the roots and tops of purple three-awn (*Aristida purpurea* [a wire grass]), side-oats grama (*Bouteloua curtipendula*), and other plains grasses are shown in Figure 10, but several other grasses and sedges also were examined.

How thoroughly and to what depths do the numerous forbs of the

Great Plains occupy the soil? Their root habits are even more variable than are those in true prairie, but one type is of special interest since it was not found eastward. Figure 11 illustrates this type of extensive root development for absorption in the surface soil. Broom snakeweed (*Gutierrezia sarothrae*), a half-shrub, occurs throughout the plains, especially in overgrazed ranges. The gumweed (*Grindelia squarrosa*), fringed sage (*Artemisia frigida*), and several other species have similar root habits. Except for 3 or 4 vertically downward penetrating roots, the roots of a cactus (*Opuntia camanchica*), which is scattered widely over the plains, spreads its fleshy roots in the surface 2 to 3 inches of soil for 4 or 5 feet on all sides of the plant. Conversely, a legume (*Psoralea tenuiflora*) has a strong taproot which frequently extends downward 1 or 2 feet before producing any branches, and very little absorption occurs in the surface soil. But maximum depths of 8 to more than 12 feet have been recorded (see Fig. 11). *Oxytropis lambertii*, a widely spread locoweed, and *Lithospermum linearifolium* have vertically penetrating but poorly branched taproots which are 8 to 10 feet long. *Yucca glauca*, small soapweed, has widely distributed lateral roots, many 20 to 30 feet long. All intermediate types occur.

In a comparison of the roots of plants of the Great Plains with those of the true prairie, two characteristics stand out rather strikingly. In true prairie, probably due to a more constant supply of water in both soil and parent materials, the roots—as a group—do not spread as widely in the surface soil. Although the depth of root penetration is marked in the Great Plains, it is usually less than in the prairie. Both the depth of penetration and the amount of root branching are profoundly affected by compact, hard soil. Plants growing on the loess hills of eastern Nebraska were more deeply rooted than those growing in glacial drift. This difference in soil type and root extent was also found later in the Great Plains; the same species of grasses and forbs were rooted much deeper in loess soils. Although some of the roots shown in Figure 11 were quite extensive, the average depth of penetration was only 6.8 feet, which is 5.7 feet less than that of the group from the true prairie (see Fig. 9). Eleven percent of the 44 species of forbs and grasses examined in the Great Plains hardlands were shallowly rooted, 28 percent had little or no provision for surface absorption, and 61 percent were both fairly deeply rooted and well adapted to absorb water—even when the surface soil was merely moist.

PLANTS IN SANDHILLS

An area of sandhills and dune topography of more than 18,000 square miles occupies central Nebraska (see Fig. 2). It lies mostly

Fig. 11.—Excavation, 20 feet deep, showing the root habits of several forbs of mixed prairie hardlands. *1, Lithospermum linearifolium; 2, Allionia linearis; 3, Artemisia frigida; 4, Chrysopsis villosa; 5, Asclepias verticillata; 6, Argemone platyceras; 7, Gutierrezia sarothrae; 8, Opuntia camanchica; 9, Oxytropis lambertii; 10, Senecio oblanceolatus; 11, Psoralea tenuiflora; 12, Lygodesmia juncea.*

north of the Platte River and has an altitude of about 4,000 feet on its western border but only 2,000 feet on the east, where it contacts true prairie. Most of the sediments that form the hills were blown from the easily eroded underlying sandy formations by strong northwest winds. The deep dune sands have little or no organic matter; in fact, little soil formation has taken place on the hills because of wind erosion. Precipitation, as in other parts of the Great Plains, varies greatly from year to year and from east to west. It averages about 17.5 inches. There is little or no runoff and the surface sand, when it becomes dry, forms an effective mulch for retarding evaporation. Normally the soil is moist to the extent of the deepest roots, which may be 12 to 20 or more feet. Studies were confined to the grasses and forbs growing on the dunes. The grasses are nearly all tall grasses, and the forbs are those found more or less abundantly only in sandy soil.

In order to compare root development more extensively under different climatic conditions, work was continued in the sandhills of Nebraska and Colorado and the root systems of 45 species of grasses and forbs were examined.

Blow-out grass (*Redfieldia flexuosa*) forms a community which represents the earliest phases of development of sandhill vegetation. It is a tall, mesophytic species which grows both on the leeward and windward sides of blowouts and on other shifting sands. It spreads by means of long, slender rhizomes which are 6 to 24 inches deep. The usually sparse and rather small clumps of grass are connected by long, coarse, tough rhizomes, which are 3 to 5 millimeters in diameter and 20 to 40 feet in length. Some of these rhizomes extend vertically upward 3 or more feet through the drifts of sand. Roots occur at the nodes of the rhizomes and may penetrate the sand to depths of about 4 feet. Roots vary greatly in length, diameter, and direction of growth, depending upon the depth and position of the rhizomes. All roots are clothed with numerous, relatively short branches, which branch again to form a dense mass of rootlets in moist sand. This grass is rarely found on stabilized soil.

Sandhill muhly (*Muhlenbergia pungens*) is a pioneer on relatively stable sands. The leafy, prostrate, rhizomatous culms are 4 to 6 inches in length. The plants grow in dense mats and produce reddish inflorescences 10 to 14 inches high. Clusters of roots arise from the short rootstocks. Most of the roots penetrate downward to depths of about 4 feet; others spread laterally 1 to 2 feet in the upper soil or grow obliquely downward. All are clothed with a dense coat of rather short laterals.

Sand reed (*Calamovilfa longifolia*) also is a perennial tall grass which

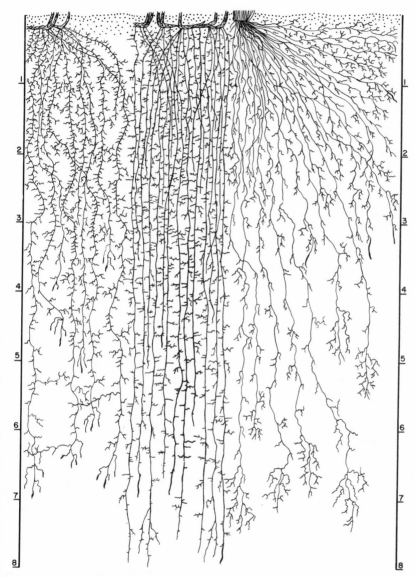

Fig. 12.—Roots of three dominant grasses of the sandhills. From left to right they are sand bluestem (*Andropogon hallii*), sand reed (*Calamovilfa longifolia*), and little bluestem (*Andropogon scoparius*). Only the parts in a one-inch strip of soil are shown.

propagates almost entirely by rhizomes, which vary from 3 to 10 inches in length. Under a good stand of sand reed there is a tangled mat of roots and rhizomes to a depth of 3 feet. Multitudes of tough, wiry roots, 1 to 2 millimeters thick, penetrated the soil in all directions, from horizontally to vertically downward. On other plants a directly down-ward course was observed. Main laterals extended outward at a depth of an inch or more to 4 or 5 feet. Branching was profuse.

Sand bluestem (*Andropogon hallii*), also a tall, coarse perennial, has an excellent root system that usually is about 6 to 10 feet in depth and extremely well furnished throughout with short, rebranched rootlets. The rhizomes are 4 to 8 inches long.

Little bluestem (*A. scoparius*) is similar in its root habit to the bluestems described in true prairie. The roots were widely spread in the surface soil, some to a distance of 3.5 feet from the base of the bunch. The wide lateral spread at greater depths and the great degree of branching is shown in Figure 12.

Blue grama (*Bouteloua gracilis*) roots penetrate 4.5 feet deep in sand. Depths of 4 to 5 feet are attained by needle-and-thread (*Stipa comata*). Sand lovegrass (*Eragrostis trichodes*), however, has a shallow, widely spreading root system. Junegrass (*Koeleria cristata*) and sun sedge (*Carex heliophila*) also are shallowly rooted.

The roots of sandhill forbs show a wide range in lateral spread, depth, and root patterns. Many are extremely well branched near the soil surface; others extend their taproots deeply, 8 to 12 feet. Root tubercles occur even at the maximum depth in certain legumes, while some species have roots with few branches. A group of 10 representative plants is shown in Figure 13.

The most extensive root systems were found on small soapweed (*Yucca glauca*) and the bush morning-glory (*Ipomoea leptophylla*). The top growth of the latter consists of a bushlike plant that is distributed over several square feet. The taproot, 6 to 24 inches thick, tapers more or less gradually to an inch or two in width at a depth of 4 or more feet. Here it may break up into numerous cordlike roots, which have been traced to depths of 10 feet and probably were much deeper. Lateral branches from the taproots spread far outward in the upper 4 feet of soil, at least 15 to 25 feet. Deeper laterals also spread far outward and grew downward, but none was greatly branched. The enlarged part of the taproot furnishes not only an enormous reservoir of food but is also a storehouse of water. This morning-glory, with purple blossoms, is scattered widely in sandy soils.

Of 45 species examined in the mixed prairie of the sandhills, only 4 may be designated as shallow-rooted, but all of the latter have a widely spreading root system. Among the 23 species with roots of intermediate depth (2 to 5 feet), only 4 have a root system which does not spread widely in the surface soil. Indeed, the widely spreading, superficial root habit is a pronounced group characteristic. Of the 18 species of

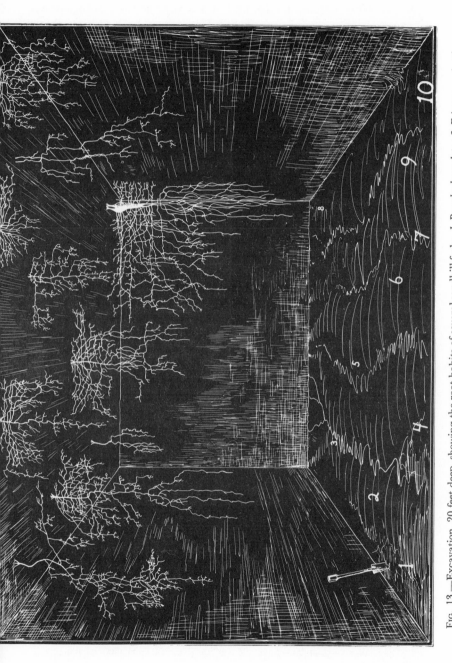

Fig. 13.—Excavation, 20 feet deep, showing the root habits of several sandhill forbs. *1, Psoralea lanceolata; 2, Eriogonum microthecum; 3, Mentzelia nuda; 4, Gilia longiflora; 5, Artemisia filifolia; 6, Petalostemon villosa; 7, Tradescantia virginiana; 8, Ipomoea leptophylla; 9, Asclepias arenaria; 10, Anogra cinerea.*

deeply rooted sandhill plants, all but 4 have widely spreading surface laterals. In the group as a whole, only 9 percent of the species have roots confined to the surface 2 feet of soil, 18 percent have few or no roots which carry on absorption in this area, and 73 percent are supplied with an absorbing system of such character as to obtain water and nutrients from both the shallow and deep soils. Many species have roots which extend into the fifth to eighth foot of soil.

The excellent development of surface-absorbing lateral roots in the sandhills is equalled only by certain species on hardlands. In a comparison of the root habits of plants in the two dry grassland habitats with those of the true prairie, two characteristics stand out rather strikingly. In true prairie species the roots, as a group, do not spread as widely in the surface soil. Also, the depth at which the largest number of absorbing roots are found is usually greater among true prairie species. This behavior is a response to a greater and more constant water supply in and below the solum.

OTHER STUDIES

Studies on root distribution in the gravel slides, half-gravel slides, and forests also were made in the mountains that border the plains. A total of 39 species of forbs and shrubs was examined (Weaver, 1919).

The 19 herbs and shrubs examined on the forest floor were relatively shallow-rooted. Almost without exception, the major portion of the absorbing system was in the surface 1.5 feet of soil.

During the course of these investigations a number of species were excavated in two or more habitats. Among ten of these, seven species showed very striking changes in their root habits (see Fig. 14); two species made almost no change; and one species exhibited only moderate differences in root development. Junegrass was examined under four distinct sets of conditions in widely separated areas. The root habits were found to be almost the same.

Two books on roots should be mentioned here since their contents will not be discussed. The first book, which concerns the root development of field crops, was reviewed as follows:

Weaver[1] has produced a book on root development of field crops that will prove of great interest and value to workers in all lines of plant science, including equally those interested in pure and applied research, as well as some practical producers. It is also sure to stimulate investigation in a field that has had far too little attention, and will interest horticulturists, agronomists, and market

[1] J. E. Weaver, *Root Development of Field Crops* (New York: McGraw-Hill Book Co., Inc., 1926).

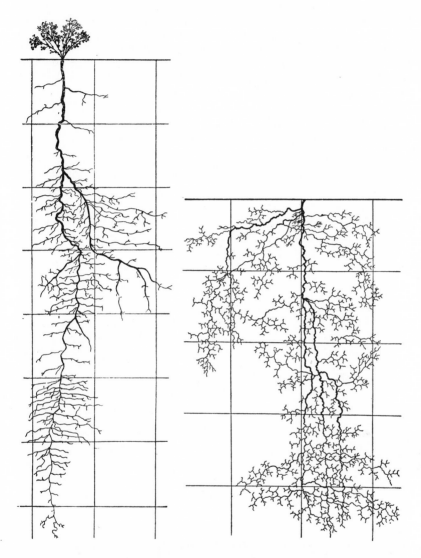

FIG. 14.—Roots of *Euphorbia montana* from the hardlands of the Great Plains and roots of the same species growing in the half-gravel slide on Pikes Peak.

gardeners in a phase of their problems that has been too little considered.

The chapter headings give a good idea of the scope and organization of the book: the environment of the root; the soil; how roots are built to perform their work; root habits in relation to crop production; root habits of native plants and how they indicate crop behavior. Chapters V–XVI deal with the root habits of wheat, rye, oats, barley, corn, sorghum, various meadow and pasture grasses, sugar beets, alfalfa, various clovers, potato, and sunflower. Chapter XVII describes the methods of studying root development. The bibliography consists of 232 citations, and there is also a comprehensive index of 16 pages.

The reviewer feels that the book is a most judicious statement of the factors affecting root development and of the effect of root development upon crop production. Many factors are considered. First is the hereditary factor. The roots of some plants are very fixed in form, being modified relatively little by environmental factors. The roots of other plants are far more plastic, being greatly modified by environmental conditions. This is generally true of crop plants. The effect of the following environmental factors are rather fully considered: soil and subsoil; water and nutrients in the soil and subsoil, including irrigation of different amounts and times of application, and the application of fertilizers of various sorts to the soil and subsoil; transplanting, tillage, intercropping; rotations of various sorts; and acids, alkalies, temperatures, and oxygen supply. The literature on the effect of conditions upon the susceptibility and resistance of roots to diseases is also brought together.

While many of the data presented are from the extensive work of the author and his associates, all other sources of information have been drawn upon, including agronomy, horticulture, physiology, and pathology. The book is an excellent critical treatise on the information in this field to date. As one reads it, he wishes that every phase of plant science as bearing on practice was as well summarized and as critically treated.[2]

The second book, *Root Development of Vegetable Crops* (also published by McGraw-Hill), illustrates and describes the root habits of 33 species of garden plants in a similar manner (Weaver and Bruner, 1927).

Plants of field and garden are often quite as deeply rooted and their lateral spread just as great as the roots and spread of native plants.

---

[2] Review by W. Crocker; reprinted by permission of the *Botanical Gazette* and the University of Chicago Press.

II.

Environment and Chief Grasses of Prairie

True prairie extends from central Texas northward into Manitoba, and Lincoln, Nebraska, lies at the midway point in this great grassland. Data from typical and extensive tracts of upland and lowland prairie at Lincoln, therefore, are representative of conditions that prevail over a wide area. The upland station was located in a tract of 180 acres of moderately rolling land three miles north of the state capital, in virgin prairie disturbed only by annual mowing, and on a level tract about 70 feet above the general level of Salt Creek. The lowland station was on a nearly level area about one-fourth mile southward.

Vegetation of the upland was dominated by little bluestem, which alone contributed about half of the cover, although some Indian grass and big bluestem were intermixed, and needlegrass and Junegrass also furnished part of the vegetation. The foliage level of grasses in summer was 12 to 14 inches; forbs were common and often somewhat taller. The lowland prairie was dominated by big bluestem, which composed about 80 percent of the vegetation, but there was also a small admixture of Indian grass, switchgrass, and nodding wild-rye. Foliage cover in late summer was 3 to 4 feet high.

PRAIRIE ENVIRONMENT

The physical factors of the environment of true prairie had been measured during the growing season for a period of 12 years, beginning in 1915. Measurements were made on both upland and lowland in extensive areas of unbroken prairie at Lincoln, Nebraska.

The topsoil or A horizon in the prairie soil (now called Brunizem) is usually about 18 inches thick; it has a granular structure. The subsoil or B horizon usually extends to a depth of about 2.5 feet; it has a higher clay content and a prismatic structure. Below the subsoil is the C horizon, composed of parent materials from which the soil develops. Since clay is less abundant, the C horizon is quite mellow. This soil is a deep, fertile, fine-textured silt loam of high water-holding capacity and is only slightly acid in reaction. It readily absorbs water, and both soil and parent materials are moist to great depths. Upland grasses penetrate from 2 to more than 5 feet, and those of the lowland penetrate approximately 3 to 10 feet. The roots are thus more extensive than the parts above ground.

The mean annual precipitation is 28 inches, of which nearly 80 percent falls during the growing season. Such a seasonal distribution of moisture is very favorable to the growth of grasses. However, periods of moderate drought are liable to occur at any time, but especially after midsummer.

Water content in the surface 6 inches of upland soil varied widely and rapidly, often 10 percent or more during a single week. It was reduced to less than 5 percent one to four times during 11 of the 12 years. Only twice during this period was the water content reduced to the nonavailable point.

Available water content in the 6- to 12-inch soil layer exceeded 5 percent three-fourths of the time but fell to 2 or 3 percent at 17 different intervals. At no time was the water available for plant growth entirely exhausted. In the second, third, and fourth foot, the water content was less variable. In general, there was a gradual decrease in the supply with the advance of the summer. This was frequently but temporarily interrupted, especially in the second foot, by heavy rains. The available supply usually ranged between 5 and 15 percent. The maximum was 21 percent, and a few times the minimum fell to 1 to 3 percent.

On the lowland, available water content was 3 to 10 percent greater in the surface foot, and often 5 to 11 percent in excess of that of the upland in the deeper soil. A close positive correlation was found between precipitation and water content, especially in the surface 6 inches.

Daytime air temperatures sometimes reached 90° F. but usually were between 75° and 85°. They were usually 10° or more higher than the temperatures at night. Soil temperatures showed a daily variation of 15° to 18° F. at 3-inch depths but only 1° to 3° at 12-inch depths. The temperature decreased regularly with depth.

The average daytime humidity varied between 50 and 80 percent during years of above-average rainfall but frequently fell to 40 to 50 percent during drier years. The average night humidity frequently was about 20 percent higher. Both readings showed weekly ranges of 8 to 20 percent. No consistent differences in humidity were found throughout the three summer months. The humidity on the low prairie usually was 5 to 10 percent greater than on the upland. Low humidity nearly always occurred during periods of low water content of soil, and usually during periods of high temperatures. Wind movement, an important factor in promoting water loss, was fairly constant, and often high.

The water content of soil and the humidity are the master factors of the environment of the prairie, and the climax vegetation is remarkably well adapted to these water relationships. That the water relationship is the controlling factor in the development of prairie is shown by studies of both distribution and growth. Although the factors of soil, temperature, wind, etc., change but little, a change in type of grassland to mixed prairie takes place farther westward with a decrease in water content and humidity. Conversely, the water relationship is the determining factor to the east in the transition from prairie to forest. The tall-grass prairie is an organic entity that exists under a certain amplitude of conditions. Its general physiognomy, ecological structure, and floristic composition remain essentially unchanged under wide variations of certain factors of the environment. Even minor changes of the water relationships, however—for example, from hilltop to midslope or lowland—are at once recorded in the type of vegetation (Weaver and Himmel, 1931).

PLANT PRODUCTION IN PRAIRIE AND PLAINS

The environment for plant growth in the Great Plains is much less favorable than that in the prairie. The two most important factors—water content of soil, as modified by runoff, and humidity—are increasingly unfavorable westward, and daily temperature changes, wind movement, and rate of evaporation also are less favorable. All of these climatic factors were continuously measured at widely spaced stations during a period of four years. A general discussion of the climate is given in *Grasslands of the Great Plains* (Weaver and Albertson, 1956). In a series of experiments, begun in 1919 and extending over a period of four years, both native vegetation and crop plants were used as a measure of environment (Weaver, 1920, 1924).

Plant production of native vegetation was ascertained by the employment of a large number of samples, each with an area of 1 square meter, at three stations: Lincoln, Nebraska; Phillipsburg, Kansas; and Burlington, Colorado (see Fig. 2). This study, begun in 1920, was continued in 1921 and 1922 to ascertain the effects of soil-moisture fluctuations on the yield from year to year. This gave an expression of the growth in each community as a unit. Since even the most uniform plant cover shows some variation in density, the samples were selected with much care and in sufficient number (about 50 at each station) to ensure dependable results. Clippings usually were made in late summer, and at approximately the same time at all stations. The clipped vegetation, when dry, was shipped to Lincoln, where it was uniformly air dried, and plant production was determined

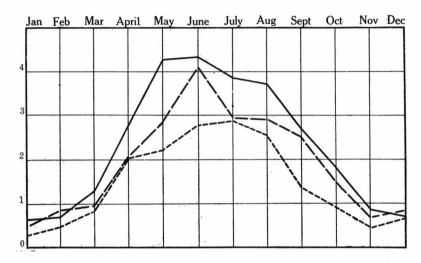

Fig. 15.—Distribution of the mean annual rainfall, in inches, at Lincoln, Neb. (solid line), Phillipsburg, Kan. (long broken lines), and Burlington, Colo. (short broken lines).

on the basis of dry weight. The average dry weight of vegetation in tons per acre is somewhat higher than that of mowed prairie since stubble is included by hand-clipping close to the soil surface.

Station	1920	1921	1922
Lincoln, Neb. . .	2.16 tons/acre	2.69 tons/acre	1.99 tons/acre
Phillipsburg, Kan. .	1.66	1.79	1.39
Burlington, Colo.. .	0.80	1.50	1.00

The average production for each year at the several stations shows a graduated series, the yield increasing with the amount and increased efficiency of rainfall (see Fig. 15). The total yield at all stations was greater in 1921 than during the preceding or following year. An examination of the record of the available soil moisture revealed the cause of the difference. It was clearly determined that the water relationships of soil and humidity of air were controlling, other factors which also were measured being merely contributory.

In another experiment, growth of winter wheat and winter rye (in 1918/1919) was ascertained at many places throughout the plains and true prairie from near the Rocky Mountains nearly to the Missouri River (Weaver, 1920). The crops were grown by farmers on non-sandy land under the usual methods of soil preparation and seeding that were common to the district in which each crop was produced. Development of the cereal crops (in feet) at the widely separated stations is shown in Table 1.

TABLE 1

GROWTH, IN FEET, OF WINTER WHEAT AND RYE AT SEVERAL STATIONS
IN GREAT PLAINS AND TRUE PRAIRIE

Station	Height of Tops (Wheat)	Depth of Roots (Wheat)	Height of Tops (Rye)	Depth of Roots (Rye)
Mixed Prairie				
Limon, Colo. (13.4*) . . .	1.8	2.8	2.3	2.0
Flagler, Colo.	1.0	1.5	2.1	2.8
Sterling, Colo. (18.1*) . .	2.0	2.8	—	—
Yuma, Colo. (20.2†) . . .	2.1	2.3	2.8	2.8
Ardmore, S.D. (16.7†) . .	2.6	4.1	—	—
Colby, Kan. (22.5†) . . .	3.2	2.3	3.5	3.6
Average	2.1	2.6	2.7	2.8
True Prairie				
Mankato, Kan.	—	—	4.2	4.7
Lincoln, Neb. (29.3†‡) . .	3.3	4.7	5.5	5.0
Lincoln, Neb. (29.3†‡) . .	3.8	6.2	6.5	5.0
Lincoln, Neb. (29.3†‡) . .	3.5	7.3	—	—
Wahoo, Neb.	3.5	5.0	3.5	5.0
Fairbury, Neb. (33.4†‡) . .	3.0	4.1	4.5	5.2
Fairbury, Neb. (33.4†‡) . .	—	—	3.9	4.2
Average	3.4	5.5	4.7	4.9

* Mean annual rainfall, in inches.
† Precipitation between July, 1918, and July, 1919.
‡ The stations near Lincoln and Fairbury were several miles apart from each other.

The development of the cereals at each station showed a striking relationship between the growth of the crop and the degree of xerophytism of the plant community. The average height of winter wheat was 1.3 feet greater in true prairie, where mid and tall grasses prevail, than in the area of mid and short grasses of the Great Plains. Also, the average depth of roots was 2.9 feet greater. Similar differences for winter rye were 2 feet for tops and 2.1 feet for root depth. As is well known, however, the height of cereals is not closely correlated with the yield of grain (see Fig. 16).

At Burlington, Colorado, due to excessive previous rainfall, the soil was quite moist to a depth of 7.5 feet and was easily molded into firm lumps with only slight pressure of the hand. The mean annual rainfall at Burlington was only 17.2 inches. Wheat was 3.5 feet tall and maximum root depth was 6 feet; both figures were far above the average. Similar occasional findings for deep, moist soil in other places in the Great Plains offer an explanation for the deep root penetration of native plants. Sometime, or at various times, during their long lifetime

Fig. 16.—Manchurian barley from a square meter of soil in experimental plots (in 1920) at Burlington, Colo. (left), Phillipsburg, Kan. (center), and Lincoln, Neb. (right). Meter rule is at right. Similar heights were attained by Marquis spring wheat and white Kherson oats.

they have been able to extend their roots deeply and have maintained this advantage.

GRASSES OF TRUE PRAIRIE

   After an intensive study of prairie vegetation over a period of ten years (1922 to 1932), a comprehensive report on the ecology and relative importance of the dominant grasses was written (Weaver and Fitzpatrick, 1932). The following account is taken from studies in prairie and pasture, during good years and extreme drought, over a period of 40 years. It deals with the life histories, distribution, and relative importance of the 10 grasses which compose about 95 percent of the grass population. Grasses of lowlands are in the left-hand column and grasses of uplands are in the right-hand column.

Big bluestem	Little bluestem
(*Andropogon gerardi*)	(*Andropogon scoparius*)
Indian grass	Needlegrass
(*Sorghastrum nutans*)	(*Stipa spartea*)

Prairie cordgrass
(*Spartina pectinata*)

Prairie dropseed
(*Sporobolus heterolepis*)

Switchgrass
(*Panicum virgatum*)

Side-oats grama
(*Bouteloua curtipendula*)

Canada wild-rye
(*Elymus canadensis*)

Junegrass
(*Koeleria cristata*)

Upland and lowland are used only as relative terms to designate soils of low and high water content, respectively.

On large, nearly level upland tracts, which lose little water by runoff, grasses requiring a good water supply often occur. These grasses have been studied in numerous prairies from South Dakota well into Kansas and from Minnesota to northern Missouri (see Figs. 17, 18). Moreover, all have been grown in experimental plots—without competition, under various degrees of competition with other species, and in very diverse habitats in prairie and pasture. They have also been closely studied in mixtures with other grasses in scores of square-meter areas over an area of 60,000 square miles (see Figs. 19, 20).

On low hillsides and everywhere on lowlands, big bluestem is the most important dominant of well-aerated soils. In saturated soils, big bluestem gives way to vast areas of prairie cordgrass. In soils of inter-mediate water content and aeration, switchgrass is the chief species.

The two bluestems are the most important dominants of the prairie; together they constitute 70 percent or more of the vegetation. They are so abundant and so vigorous that their influence upon the habitat and their effects upon other species to a large degree determine the con-ditions under which all the remaining species associated with them must develop. Big bluestem is more mesic than little bluestem and is best developed on lower slopes and on well-aerated lowlands. A cover of big bluestem of 80 to 90 percent is common on well-watered areas. On the lower and middle slopes of hills it shares dominance with little bluestem, and occurs in amounts of 5 to 25 percent on all but the driest hilltops.

Seedlings of big bluestem develop rapidly and, in the absence of competition, clumps 12 to 18 inches tall may occur at the end of a single growing season. During the first summer, even under competi-tion, the root system may reach a depth of 2 to 4 feet. In the lowland sod, roots from stem bases and rhizomes have a vertically downward course, but on uplands, among bunch grasses, many roots spread widely in the surface 6 to 12 inches of soil. Tillers begin to appear about seven weeks after germination and the seedlings develop a small tuft or bunch. The tillering habit is pronounced, as is the production of rhizomes;

Fig. 17.—Loess hills near Ponca, in northeastern Nebraska, covered with prairie. The dominant grasses are *Andropogon scoparius*, *Stipa spartea*, and *A. gerardi*. Photo was taken June 7, 1932.

Fig. 18.—Lowland prairie in southeastern Nebraska after hay has been mowed and stacked.

Fig. 19.—An Iowa prairie near Atlantic.

Fig. 20.—Extensive, 300-acre prairie near Lincoln, Neb. Photo was taken in August, 1943.

hence this grass becomes a successful competitor among other species. Moreover, the seedlings are very tolerant of shade.

The Andropogons are of southern derivation and usually do not renew their growth until mid-April; this is several weeks after Kentucky bluegrass, which is an invader, begins its growth. Indeed, bluegrass may blossom before the plant is much shaded. *Poa pratensis* was found in about 80 percent of the big bluestem prairies. Removal of the bluestem by mowing is distinctly advantageous for the growth of bluegrass, both in fall and early spring. Development of this perennial bluestem, especially on lowlands, is very rapid. By June 1, the foliage height is 14 to 16 inches, and by July 1 it is 2.5 to 3 feet. The leaves spread outward so widely that the top of a clump usually occupies 1.5 to 2 times the area of the base. Flower stalks begin to appear above the general level of the foliage early in July but anthesis does not reach its maximum until August or September. Then the rather woody flower stalks, with their usual 15 or more terminal or axillary racemes, attain a height of 7 to 10 feet (Fig. 21). Each seed is provided with a long, bent, twisted awn.

Big bluestem, like its tall-grass associates, is an excellent grass for pasture. Liked by all classes of livestock, it also is one of the chief constituents of wild or prairie hay. Usually two cuttings are made each year for hay.

Indian grass, a chief associate of big bluestem, is of much less importance. It is a coarse, perennial species whose habitat requirements are almost identical with those of big bluestem. It occurs in greatest abundance in Kansas and farther southward, but it is widely distributed throughout the prairie. It may form 90 percent of the vegetation where local stands occur in ravines, and 5 to 20 percent where it is best developed on lowlands. The more usual percentages, however, are 1 to 5 percent, and in many lowlands Indian grass is almost totally absent. In big bluestem sod, its occurrence as isolated stems or very small clumps is usual. Where occasional flooding or burning occurs, Indian grass increases greatly in abundance. It has a more erect growth than big bluestem, usually has broader leaves, is slightly lighter green in color, and because it has very prominent ligules it can be readily distinguished from big bluestem. Moreover, the large panicles that terminate the 6- to 9-foot flower stalks are very different from the fingerlike racemes of the bluestem. Like the big bluestem, each seed of Indian grass is furnished with a long, bent, twisted awn. The seeds are rather large, and usually viable, and they germinate readily unless they are buried more than a half-inch deep. The vigorous seedlings endure a wider range of extremes as regards

Fig. 21.—A fine clump of *Andropogon gerardi*, about 6 feet tall, and a big bluestem prairie in autumn.

drought than most lowland grasses. This may explain, in part, the habit of this species of readily invading disturbed areas. Seedlings develop rapidly. Under severe competition, however, tillering is almost nil, which may account for the fact that single stems of this grass are scattered over the prairie.

Little bluestem is a bunch grass which forms the great bulk of the cover of uplands. In abundance, it easily exceeds all of the other upland species combined. In the north and west it is often accompanied by needlegrass (*Stipa spartea*), and sometimes—on drier soil—is locally replaced by it. This bluestem forms a much-interrupted sod northward, with mats, tufts, and bunches so dense that other species usually occur

Fig. 22.—Little bluestem pasture near Belleville, Kan.

only in the interspaces. But where slopes are steep and runoff is accelerated, the bunch habit becomes pronounced. On the deep soils of the steep, drier loess hills, little bluestem alone frequently composes 90 percent of the vegetation. In drier soils as a whole, including level uplands, it composes 50 to 75 percent.

Germination is often low but the seedlings are vigorous. During the first summer they attain a height of 6 or more inches, and tiller profusely. The fine and extremely well-branched roots reach a depth of 24 inches or more. In the absence of competition it may complete its life cycle by producing flower stalks and seed the first year, but in the prairie this may require two growing seasons, and ordinarily requires three or more growing seasons. Mature plants are about 15 inches tall, lateral spread of roots is 12 to 18 inches, and root depth is about 5 feet. On lowland, in competition with the tall grasses, this mid grass may attain a height of 2 to 3.5 feet.

The leafy stems grow compactly in the bunch or mat; 100 to 300 stems often are crowded into a circular bunch 4 inches in diameter (Fig. 22). Many stems develop several tillers, and all are leafy to the base. The density of the bunch or mat varies with the habitat and also with age. Both bluestems are very long-lived. Deterioration of the bunch in mowed prairie nearly always occurs first in the center of the clump and proceeds toward the periphery. The size of the sod mat varies greatly, from solid mats 1.5 by 2 feet in extent to mats of 6 to 8 inches.

Growth begins in late spring, at a time when Junegrass and needle-grass (species of northern extraction) have well-developed bunches. Flower stalks appear by mid-July and are abundant late in August. The inflorescence is terminal, both from the main stalk and its branches. The flower stalks are of variable length, 1.5 to 3 or more feet. The stalks are thickly grouped, and in late fall the whole top of the plant resembles a man's gray beard, as the words *andro* ("man") and *pogon* ("beard") indicate. Seeds are produced in great abundance; each has a twisted awn about a half-inch long. The beautiful coloration of the prairie in autumn is largely due to the various shades of yellow, red, and bronze afforded by the drying plants of the bluestems and Indian grass. When the dry stems do not interfere, little bluestem is a rather good grass both for grazing and hay.

Needlegrass (*Stipa spartea*) is a perennial, cool-season bunch grass of boreal origin. The usually circular bunches are often only 1.5 to 2.5 inches in diameter but, because of extensive peripheral tillering, they may attain a diameter of 4 to 5 inches. In mature stands the bunches usually are rather widely spaced and more or less intermixed with other grasses, especially Junegrass and little bluestem. In spring, growth is very rapid, and a height of 2 feet may be attained by the first of June. Even in early summer the long slender leaves, shiny beneath, have attenuated ends and dry, dead tips. Foliage height is about the same as that of little bluestem, and, like little bluestem, it depends upon the degree of competition and the amount of available water.

Flower stalks usually are 3 to 4 feet tall (Fig. 23). Flowering follows quickly, and by June 10 the twisting of the 4- to 6-inch awns indicates the ripening of the seed. The "seeds," as in most grasses, are really fruits and may be blown a considerable distance by the wind, especially when the long awns twist together in clumps. Unlike most other grasses, the seed germinates best when it is planted deep. It is buried 1 to 3 inches deep in the soil by the torsion of the bent and twisted awn, which is held from turning by the bases of grass stems. Even after the seeds have fallen the persistent, broad, drying scales on the spikelets glisten in the sun so that a whole hillside—if the plants are at all abundant—has a silvery appearance. Wherever needlegrass is thick (and it dominates in many places), the ground cover is very open, and the greater the apparent density of this grass the barer the soil. Although vertically descending roots usually are about 5 feet deep, the roots of needlegrass spread widely and thread the upper soil.

Fig. 23.—A bunch of *Stipa spartea*, about 3.5 feet tall, after the seeds had fallen (right), and a bunch which was retarded by severe drought.

The early growth of needlegrass, before most other plants in grazed prairies, is detrimental to the species. It is so palatable to livestock that it is frequently exterminated by grazing animals. Elsewhere, it should not be mowed for hay until after the seeds have fallen.

Prairie dropseed (*Sporobolus heterolepis*) is a bunch grass. The bunches usually are 4 to 7 inches in diameter, but larger ones—up to 15 to 18 inches—also occur (Fig. 24). In vigorous bunches, under high rainfall, the stems are densely aggregated in the bunch, but where rainfall is light the number is often 25 to 50 percent less. The best development of this grass occurs on hilltops and dry upper slopes, where it may dominate local areas by forming 80 percent or more of the vegetation. It is often intermixed with little bluestem and needlegrass. The leaves do not stand erect but curve over gracefully, so that the top of the bunch usually has a width about 3 times as great as that of the base. The shining, light-green color and the attenuated leaves with dead tips enable one to distinguish the species from little bluestem. This dropseed attains about the same height as little bluestem. Many roots are widely spread in the surface soil and others are about the same depth as those of the bluestem, 5 feet. It has numerous short, woody rhizomes, and stems with swollen bases, and—unlike most grasses—it retains its abundant leaves in winter. The leaves spread outward and downward and much debris collects between the bunches of unmown plants (see Fig. 25). Flower stalks of this warm-season perennial appear in midsummer, but anthesis reaches its height late in August or September. The panicles are open and loose. The rather large seeds ripen in September and soon fall to the ground. During spring and summer this grass furnishes excellent grazing forage.

Fig. 24.—A field of prairie dropseed (*Sporobolus heterolepis*).

Fig. 25.—Bunches of unmowed prairie dropseed several years old: with dead leaves removed (left) and with old leaves intact (right).

Junegrass (*Koeleria cristata*) is a bunch grass of smaller stature than any of the preceding upland grasses. It is of boreal origin and is most abundant in the northern prairies. Southward, it may be scarcely if at all represented, but it usually forms 1 to 3 percent of the cover, at least locally. Except in disturbed places, an abundance of more than 5 to 10 percent is rare. Although a perennial, it is not long-lived and depends mostly upon reseeding. Blossoming begins late in May and reaches its height by the middle of June. Seeds are produced in great abundance but the seedlings usually have difficulty in becoming established. During summer drought the mortality is always high, and many seedlings are winter-killed. Even the root system of mature plants is relatively short.

Side-oats grama (*Bouteloua curtipendula*) is another grass that is scattered widely throughout the prairie. It occurs in all types of situations but rarely in great abundance. It is a very drought-resistant, warm-season, perennial mid grass which is dominant on the Great Plains. In true prairie, however, where it is usually found on steep banks of ravines, on dry ridges, and in slightly disturbed places, it may form 10 to 60 percent of the plant cover. It ranges widely from the big bluestem community throughout all types of upland. Its stature is about the same as that of other upland grasses (Fig. 26), but its roots penetrate somewhat more deeply. When the prairies are damaged by extreme drought or by grazing, this grass rapidly increases in abundance despite its high palatability for cattle.

Prairie cordgrass (*Spartina pectinata*) formerly grew over hundreds of square miles on the first bottomlands of the Missouri River and its tributaries, and on soils too wet and too poorly aerated for the development of big bluestem or switchgrass. Typically, it grows on the edge of sluggish streams or ponds and in waterlogged or wet soil. Vast areas of the first bottomlands were covered with this grass, often in almost pure stands. On the wetter side, hydric species of tall sedges and rushes

Fig. 26.—Side-oats grama (*Bouteloua curtipendula*), about 2.5 feet tall.

are plentiful; on the mesic side, these give way to big bluestem, usually through a transitional zone that is characterized by *Panicum virgatum* and *Elymus canadensis.*

The seeds germinate readily in wet soil and the seedlings develop rapidly. They may attain a height of 2.5 feet by the end of their first growing season. The seedlings, however, are not very tolerant of shade, and reproduction, except in bare areas, is undoubtedly accomplished almost entirely by rhizomes and tillers. Established plants renew their activities rather early in April and growth is more rapid than that of other grasses. In marginal areas, where prairie cordgrass is intermixed with switchgrass or big bluestem, it soon overtops them. Foliage height is usually 5 to 7 feet and mature leaves are often half an inch wide and 2.5 to 5.5 feet long. Flower stalks of this coarse-stemmed grass range from 5 to 10 feet in height. The panicles are very large and conspicuous, and usually consist of 10 to 20 spikes.

Prairie cordgrass makes good hay if it is cut before the woody stems are much developed. A common practice is to mow it three times each year. Renewed growth is prompt because much food is stored in the complex mass of coarse rhizomes.

Panicum virgatum, a tall, coarse, sod-forming grass, is an important dominant of low, moist soil (see Fig. 27). It occurs in dry portions of the prairie cordgrass community that are too poorly aerated for the growth of big bluestem. It is rarely found in extensive pure stands and its clumps and patches usually alternate with prairie cordgrass and tall sedges. It almost invariably occurs as a transitional species along ravines and draws wherever big bluestem gives way to prairie cordgrass. It is perhaps the most mesic grass of the true prairie. Its rank growth at the bottom of ravines, where it is often 5 feet high, gradually decreases to about 3 feet near the top.

The large seeds, although produced abundantly, do not germinate well until they have undergone a period of dormancy. Then, when planted, germination is prompt and development both below and above ground is rapid. It is a warm-season grass and renews its growth late in April. Unless competition is too severe, tillering and production of rhizomes begin in 5 to 7 weeks. Three months after germination, heights of 12 to 20 inches are attained and the roots are 12 to 30 inches deep. This species does not tiller as rapidly or abundantly as big bluestem. In addition, winter-killing of seedlings often is very great.

Mature plants attain a height of 4 to 7 feet on lowlands by the middle of July; then large, open, spreading panicles begin to appear. These are 12 to 20 inches long and 16 to 20 inches wide. Flowering begins in July and continues until frost. The seeds are shed in late fall

Fig. 27.—Switchgrass (*Panicum virgatum*) about 7 feet tall and in seed in mid September.

or winter. This grass, when young, affords good grazing, but as the stems become lignified the forage value decreases. It also produces large yields of hay.

Canada wild-rye (*Elymus canadensis*) is a widely distributed prairie grass but is of much lower rank than any of the preceding lowland species. It is a tall, coarse grass, of a high water requirement, that exhibits its best development in the wet areas between prairie cordgrass and big bluestem but intermingles with both of these communities. Elymus occurs only in isolated clumps in uplands. Even on lowlands it forms pure stands only in very local areas (see Fig. 28).

The rather large seeds of this cool-season grass show a high rate of germination and the seedlings are not easily winter-killed. Tillering

Fig. 28.—Canada wild-rye (*Elymus canadensis*) in fruit in late summer.

begins 4 or 5 weeks after germination. The short rhizomes result in the formation of bunches or clumps of stems. Mature plants renew their growth 15 to 20 days earlier than the warm-season big bluestem. Foliage heights of 2.5 to 3.5 feet are attained by the middle of June and flowering occurs in midsummer. The broad spikes are 6 to 9 inches long. The spikes bend over gracefully during the ripening of the seed, which accounts for the very appropriate name, nodding wild-rye. Elymus is a good forage grass when it is harvested early. If it is left to mature in the field, not only do the stems become lignified but the heads often become infested with ergot, which is harmful to livestock.

In ecological study it is generally accepted that the important plants, especially the dominants, deserve most careful study because they markedly reflect the impress of climate and react strongly to the forces of the habitat.

III.

The Prairie

This study, begun in 1929, treats of the structure of vegetation in the several types of prairie, the secondary grasses, the ecology of the forbs, and other features of grassland. It resulted from five years of intimate association with the great grasslands of North America. The investigations were made to clarify some of the many problems presented by this vast natural unit of vegetation, to better understand the importance and significance of grassland and its utilization, and to furnish a permanent record of a rapidly vanishing vegetation. The data presented here are taken from a 186-page monograph (Weaver and Fitzpatrick, 1934).

INTRODUCTION TO THE PRAIRIE

The prairie appears almost monotonous in the general uniformity of its plant cover. Its main features are the absence of trees, the scarcity of shrubs, the dominance of grasses, and a characteristic xeric flora. Neither geological formation, topography, nor soil determines the character of the flora which develops under the master hand of climate.

The area investigated lies in six states. It includes the grasslands of the western third of Iowa and of the eastern third of Nebraska. On the south it extends into Missouri and Kansas to the Kansas River; northward it includes a small part of southwestern Minnesota and a larger part of southeastern South Dakota. Forest is represented only by narrow belts of woodland along the streams and in an area 2 to 15 miles wide along the Missouri River. In this vast tract of more than 60,000 square miles, 135 representative areas, each 20 to 360 acres in extent, were selected and carefully studied. They were chosen to represent typical topography and to obtain, so far as possible, a rather uniform distribution throughout the region (see Fig. 29). The results recorded here are believed to apply, in the main, to a much larger area.

This portion of the prairie grows under a climate that is characterized by moderately long, cold winters and by a long growing season with hot summers. The growing season at Lincoln, Nebraska, includes 160 to 170 days without severe frost. Awakening of plants usually

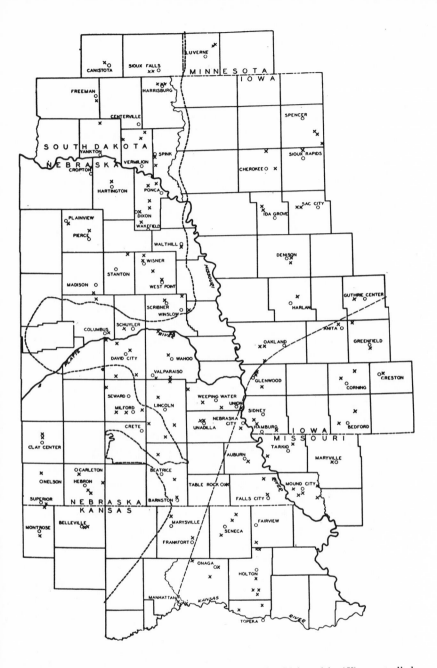

Fig. 29.—Outline map of portions of the six states in which prairies (X) were studied. The subdivisions are counties and the circles indicate towns and cities. The broken line across the southeastern part indicates the isohyet of 30 inches; the broken line running from Minnesota to Kansas separates the Prairyerths (on the east) from the Blackerths.

begins late in March and continues until mid-April, although some plants continue activity until late in October. Mean annual precipitation in the southern South Dakota portion is about 25 inches; in Missouri the rate is about 36 inches. Its distribution is of the Great Plains type: about 77 percent occurs between April 1 and September 30, and about 14 inches fall during May, June, and July. (The prairie environment and soil have been explained more fully in preceding pages.) The region's cover of grassland is fairly similar throughout in species, structure, and manner of growth. The various kinds of prairie are clearly related to the water content of the soil, but only obscurely—if at all—to local soil types.

Within the prairie the conditions of life are severe. Though the soil is rich and deep, water frequently is scarce and the plants that share it are legion. Deficiency of water usually occurs when the air is driest, when the temperature is high, and when the prairie is swept by desiccating winds. So numerous are the species that those of greater stature shade the shorter ones, often to the extent that seedlings and lower leaves die for lack of food. After thousands of years, the species have adjusted to the environment. The plants, with few exceptions, are remarkably free from disease, regardless of the weather, and are little injured by high winds or extreme heat. They may be harmed by late freezing or—infrequently—be stripped of their leaves and battered to the ground by hail, but they rarely or never are killed. Those that were unfit have disappeared and those that remain have reacted to the factors of the environment so thoroughly that, as species, they successfully meet the most severe conditions.

The problem of an adequate water supply has been met by the development of deeply penetrating and thoroughly efficient root systems. Moreover, roots occur in layers: shallow, medium, and deep. The underground plant parts—rhizomes, roots, bulbs, etc.—are storehouses of food during the long period of winter dormancy and account for the rapid growth of the plants after their early awakening in spring. The perennial life-habit is exhibited by all of the dominant species, as well as by most of the species that are of secondary rank. Reproduction is largely vegetative, by tillers, rhizomes, root sprouts, or the like. Seedling survivors are very few. Roots and rhizomes form so dense a network in the sod that invaders—even in apparently bare areas—cannot compete successfully with the prairie population. Weeds are excluded.

The beauty and quiet calm of the grassland should not obscure the fact that the prairie is both a field of battle that is centuries old and a field in which the conflicting species, never wholly victorious nor entirely vanquished, each year renew the struggle. There is a bitter struggle for mere existence—for light, water, and nutrients—all of which are eagerly sought by numerous competitors. As a result, the population becomes enormously overcrowded for the best development of the individual species, which consequently are reduced in size and underdeveloped in comparison to the stature they could attain. They often fruit sparingly, rather than abundantly, and require years to accomplish what, if unhindered by their fellows, might have been accomplished in a single season.

So many species—often a total of 200 or more per square mile—can exist together only by sharing the soil at different levels, by obtaining light at different heights, and by making maximum demands for water, nutrients, and light at different seasons of the year. Legumes add nitrogen to the soil; tall plants protect the lower ones from the heating and drying effects of full insolation; and the mat-formers and other prostrate species further reduce water loss by covering the soil's surface, living in an atmosphere that is much better supplied with moisture than are the windswept plants above them. Light is absorbed at many levels; the more-or-less-vertical leaves of the dominant grasses permit light to filter between them as the sun swings across the heaven. Compared with man's crops of wheat or maize (in adjacent fields on virgin soil), fluctuations in temperature of both soil and air are much less in prairie, humidity is consistently higher, and evaporation is decreased. The prairie's demands for water and light increase more gradually and extend over a longer period of time. Less water is lost by runoff or by surface evaporation. When drought occurs, the vegetation gradually adjusts itself during a period of stress (see Figs. 30, 31).

As one travels over the rolling hills and across the intervening valleys, the orderly sequence of the dominance of one species being replaced by that of another recurs again and again, as often as the changes in habitat warrant. Many forbs of early spring always remain near the surface of the soil. Certain species for a long time avoid competition with the grasses by producing an elongated stem before they unfold their leaves; thus an abundance of light is assured. Still other species—and these are the real competitors of the grasses—demand leaf space from soil surface to leafy top, which often extends to a greater height than that of the grasses.

PURPOSE OF THE STUDY

The purpose of this study was to determine the nature, development, continuity, and intimate structure of a representative section of

FIG. 30.—Prairie, near Nebraska City, with an intermixture of needlegrass and bluestems.

FIG. 31.—Lowland prairie on the flood plain of the Platte River, near Columbus, Neb.

the most extensive and perhaps the most highly differentiated of plant climaxes, the Grassland Formation of North America. The study is based upon the examination of 135 different tracts of prairie that extend throughout the region. These prairies were mowed annually, in fall—some over a period of 50 years—and various points of much importance should be emphasized. The prairies occurred on all types of topography and soil: some occupied steep, glaciated hills, and others were on hills of loess. Considerable areas of broad "first" and "second" bottomlands were still in grassland, as well as great areas of almost level uplands. Moreover, these prairies were unmodified by invaders from the surrounding, cultivated vegetation.

The methods of study combined both the extensive and intensive types, and the studies were undertaken only after the writers, who had spent many years in the grassland, were well acquainted with almost all the species they encountered (Weaver and Fitzpatrick, 1934). Indeed, preliminary studies were made so that the plants could be recognized in their vegetative stages of development. The investigators thus were left free to give their major attention to the ecological rather than the taxonomic phases of the problem. Major emphasis was placed upon the more important species, as regards structure and relatively little attention was given to problems pertaining to distribution and variations of rarer constituents of the flora. This circumstance was fortunate because the ecological problems alone offered a very complex field for study.

Although many of the prairies were visited several times during the growing season, and perhaps half of them at least twice, others were studied at only one period. In fact, several fine tracts were found broken, and others fenced and grazed, when they were visited the second year, thus portending the final destruction of the native grasslands. The actual quadrating and listing of forbs was done mainly between June 10 and mid-August, mostly during a period of three years (1929–1931), and a fourth summer was used as a check.

COMPOSITION AND STRUCTURE

Upon entering a prairie the investigators' first problem was to gain a preliminary knowledge of the type or types (communities) it contained, their extent, and their relative importance. Each type of vegetation was studied in detail and transitions from one to the other were carefully noted. The surveys were not superficial: hundreds of yards were traversed on hands and knees and the surface of the plant cover was penetrated to determine its constitution and form of construction—a most interesting and fascinating task. It was fortunate that the development, distribution, and interrelationships of most of the dominants' and subdominants' parts below ground were already well understood.

The distribution of grasses in the several types or communities of the grassland was ascertained. Hundreds of samples of the vegetation in the six states, each consisting of 1 square meter, were taken in the several types of grassland. A typical chart quadrat is shown in Figure 32. These were made only at the expense of considerable time. List quadrats were nearly always employed to ascertain the percentage of each species in the sample. Their significance is greatly enhanced by the fact that the data disclosed by their analyses were confirmed by literally hundreds of critical and detailed observations in the 135 prairies in which these studies were made.

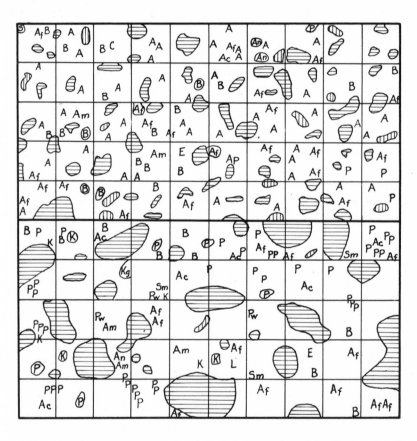

Fig. 32.—A square-meter quadrat in a little bluestem prairie. The lower half shows the bunch type; the upper half shows the sod-mat type. *A* or horizontal hatch = *Andropogon scoparius*.

Typical variations in the plant cover that were due to the water content of the soil and the humidity of the air were observed on many hillsides. For brevity, only the most abundant species are given and the data are those obtained in the field. Table 2 shows the dominance of little bluestem on a hillside in Iowa and its decrease and final replacement by big bluestem. Table 3 is an abstract of data from a Nebraska prairie, and Table 4 is an abstract of data from a Kansas prairie. The general findings in many other prairies were similar. On nearly level lowlands, dominance of the tall grasses almost always occurred. Transitions between the several lowland types of grassland are little affected by topography, except as it determines the accumulation of runoff water and the consequent soil aeration and the depth of the water table.

TABLE 2

QUADRATS, SHOWING THE VEGETATION PERCENTAGES, ON A NORTH SLOPE NEAR HARLAN AND ON A SOUTH SLOPE NEAR GUTHRIE CENTER, IA. (JUNE 15, 1930)

Species	Hill Crest	Midslope		Lower Slope	Midslope		Lower Slope	
Andropogon scoparius	90	57	..	..	89	90	48	46
Andropogon gerardi	5	33	94	95	3	5	47	48
Poa pratensis	2	2	5	3	2	..	2	5

TABLE 3

SERIES OF PERCENTAGE QUADRATS ON A LONG WEST SLOPE AT WISNER AND ON A STEEP NORTH SLOPE AT VALPARAISO, NEB. (AUG. 1, 1930)

Species	Upper Slope				Midslope			Lower Slope				Hilltop	Upper Slope	Lower Slope	
Andropogon scoparius	47	62	38	32	..	..	..	..	..	..	..	61	50	..	..
Andropogon gerardi	10	15	8	48	75	74	77	89	92	86	89	9	37	86	97
Poa pratensis	12	12	30	10	15	14	15	5	3	10	5	3	5	10	2
Bouteloua gracilis	..	..	..	..	..	..	..	..	..	..	..	20	..	..	..

TABLE 4

SERIES OF PERCENTAGE QUADRATS, FROM NEAR THE TOP TO THE BASE OF A LONG, RATHER GENTLE SOUTH SLOPE, NEAR TOPEKA, KAN. (JULY 19, 1930)

Species	Upper Slope				Midslope		Lower Slope		Level Base	
Andropogon scoparius	65	40	81	69	52	75	15	18	16	5
Andropogon gerardi	10	9	10	8	30	20	63	60	80	89
Sporobolus heterolepis	15	50	..	..	6	..	..	..	..	..
Panicum virgatum	..	..	2	8	10	4	..	5	2	..

Table 5 reveals that little bluestem alone furnished an average basal cover of 55 percent in 180 quadrats throughout the area. It also composed 90 to 98 percent of the basal cover in several unit areas, and in a third of the quadrats it exceeded 70 percent. The two bluestems together constituted 80 percent of the entire plant cover. The wide distribution of bluegrass, which occurred since settlement, should be noted, as well as its percentage of cover.

The distribution of needlegrass was much more regular than that of prairie dropseed; it occurred in twice as many quadrats. The wide distribution of Indian grass is shown by the fact that it was present in more than half of the quadrats, just as its sparseness is illustrated by its small proportion (1.8 percent) of the plant cover. The basal cover of forbs is small and their role in grassland is considered elsewhere. Many minor grasses and sedges completed the total cover of vegetation.

TABLE 5

PERCENTAGE OF BASAL COVER IN "ANDROPOGON SCOPARIUS"
TYPE, COMPOSED OF VARIOUS GRASSES AND FORBS, AND
PERCENTAGE OF 180 QUADRATS IN WHICH EACH OCCURRED

Species	Cover %	Occurrence %
Andropogon scoparius	55.0	98
Andropogon gerardi	24.8	99
Poa pratensis	4.7	80
Stipa spartea	2.5	40
Sporobolus heterolepis	2.7	20
Sorghastrum nutans	1.8	51
Bouteloua curtipendula	0.6	32
Panicum scribnerianum,		
P. wilcoxianum	0.4	36
Koeleria cristata	0.6	34
Elymus canadensis	0.0	4
Panicum virgatum	1.3	14
Forbs	4.1	90
Total	98.5%	

A number of important facts are revealed by an analysis of the data in Table 6, which is based upon 155 quadrats throughout the area. Big bluestem was represented in every quadrat and, on the average, composed 78 percent of the vegetation. Conversely, the small amount of little bluestem is revealed. Kentucky bluegrass was more abundant here, and slightly more widely distributed than on uplands. The role of Indian grass was still a minor one. Only very small amounts of the usual upland grasses were found. Minor grasses and sedges, which are not shown in the table, totaled 1 percent.

TABLE 6

PERCENTAGE OF BASAL COVER IN "ANDROPOGON GERARDI"
COMPOSED OF VARIOUS GRASSES AND FORBS, AND PERCENT-
AGE OF 155 QUADRATS IN WHICH EACH OCCURRED

Species	Cover %	Occurrence %
Andropogon scoparius	2.0	19
Andropogon gerardi	78.0	100
Poa pratensis	8.8	88
Stipa spartea	1.9	31
Sporobolus heterolepis	0.1	1
Sorghastrum nutans	1.9	37
Bouteloua curtipendula	0.1	7
Panicum scribnerianum,		
P. wilcoxianum	0.3	28
Koeleria cristata	0.1	10
Elymus canadensis	0.1	12
Panicum virgatum	1.7	22
Spartina pectinata	0.4	12
Forbs	3.6	74
Total	99.0%	

A series of quadrats from the much-less-extensive needlegrass community revealed that the dominant alone composed half of the vegetation. The two bluestems shared second rank equally, 17.5 per-cent each. Bluegrass and Junegrass—5.1 and 1.6 percent, respectively—were next in abundance.

Distribution of the major grasses in the bluestem types was uniform throughout the area. Hill crests and dry slopes were dominated by little bluestem, which furnished 40 to 90 percent of the basal cover. On moist midslopes, dominance was often shared equally by the blue-stems. On lower slopes and in ravines, big bluestem dominated, and on well-drained lowlands it furnished 85 to 88 percent of the cover.

The average basal cover by the little bluestem type was 15.3 percent, which varied little from year to year. Foliage cover varied from 55 to 100 percent. Basal cover increased 5 percent in areas of greatest rainfall (precipitation 32 to 36 inches) over areas of lighter rainfall (25 to 32 inches). This resulted from a decrease of little blue-stem and other xeric grasses and an increase of big bluestem and other mesic ones. The average basal area in the big bluestem type was 13.3 percent and the foliage cover usually was 90 to 100 percent. Basal cover increased 9 percent in the wetter area over that in the drier area due to the decrease in abundance of several xeric grasses. Average basal cover in the needlegrass community was 10.9 percent.

The foliage of little bluestem was 5 to 9 inches higher in the southeast than in the drier southwest or northwest. Foliage of big blue-stem showed differences in height of 6 to 12 inches. Bunches of little bluestem and other grasses were more completely filled with stems southeastward. Flower stalks were larger and more abundant, occurring even during dry years and yields of hay were greater.

STUDY OF FORBS

In each prairie the relative abundance and importance of each species other than the grasses also were studied. Their relative ranking was determined only after ascertaining the abundance, size, duration, density, gregariousness, and basal and foliage cover of each species. The criterion for ranking each of 75 species of upland forbs and 67 lowland forbs was the effect of the forbs upon the cover of grasses and the portion of the ground and foliage cover that they actually occupied. This involved a long, continuous study.

Forbs that occurred in great abundance and were of considerable importance in any particular prairie were given a ranking of 1, a society of the first class. They were widely, but not necessarily continuously, distributed. Forbs of great abundance and much importance, but not sufficiently so to be ranked as a society of the first class, were listed in a second grouping, class 2. On the other extreme, forbs that were so rare that they were found in only one or a few places were given a ranking of 5 for that particular prairie. Where they were of somewhat more frequent occurrence and importance, the ranking was raised to 4. A large number of forbs that were not sufficiently important to be given second rank were found in almost every prairie, but they were of too great abundance and occupied so much space in the plant cover that they were assigned an intermediate ranking of class 3. Thus, while the ranking of a particular species varied from prairie to prairie, such a census of the population revealed the species which were of greatest importance in most of the prairies over the area as a whole, which were only locally important, and which were consistently present but only in small numbers.

The names of all plants were listed when they were first observed by an investigator (three men—the two writers and an advanced graduate student—walked back and forth across a prairie a few yards apart), but the groupings were made tentatively for each prairie only after the general survey was well under way. When it was completed, the rankings were changed or verified, as further study warranted. When a disagreement arose, it was settled only after further examination.

A record of the vegetation also was obtained by means of more than 120 photographs. A curtained background was often used to still the breeze, to separate an individual plant from other vegetation, and to obtain greater definition of its parts.

TABLE 7*

DISTRIBUTION AND RELATIVE IMPORTANCE OF THE SEVEN CHIEF FORBS OF UPLAND
PRAIRIE AND THEIR OCCURRENCE IN LOWLANDS

Species	High Prairie						Low Prairie					
	1	2	3	4	5	0	1	2	3	4	5	0
Amorpha canescens.	74	6	7	4	2	7	9	0	9	25	9	48
Helianthus rigidus.	66	7	7	4	3	13	19	5	14	11	0	51
Aster ericoides	44	15	21	5	0	15	25	3	22	14	11	25
Antennaria campestris	39	20	13	4	2	22	8	6	11	11	8	56
Erigeron ramosus	47	11	18	6	3	15	15	15	10	17	15	28
Solidago glaberrima	32	12	20	12	0	24	9	6	9	17	3	56
Psoralea argophylla	37	6	13	10	8	26	8	3	8	10	14	57

*The numbers unders columns *1, 2, 3*, etc., give the percentage occurrences as societies of the first, second, etc., class, and the numbers under column *0* give the percent of prairies in which the species did not occur.

A table was made of a group of the 75 most important species of upland forbs, arranged in the order of their rank as determined by their percentage of occurrence as societies of the first and second class. A portion of the table is given in Table 7. Thus *Amorpha canescens* was absent from only 7 percent of the upland prairies and was ranked as a society of the first class in 74 percent of the upland prairies, as a society of the second class in 6 percent of the upland prairies, and so on.

Amorpha canescens, or lead plant, a half-shrub that behaves as an herb under annual mowing, is the most important (Fig. 33). This

FIG. 33.—Lead plant (*Amorpha canescens*) [left]) and prairie rose (*Rosa suffulta*).

shade-enduring legume is leafy to the base of the stems, which attain a height of 1.5 to 3 feet. The widely spreading roots absorb to depths of 6 to 16 feet. The six species which follow all occurred in at least three-fourths of the uplands and ranked as societies of the first class in one-third or more. Two important species are illustrated in Figure 34.

The next high-ranking species, in order (beginning with the left-hand column), are:

Petalostemon spp.	*Rosa arkansana*
Echinacea pallida	*Coreopsis palmata*
Euphorbia corollata	*Kuhnia eupatorioides*
Solidago rigida	*Psoralea floribunda*
Astragalus crassicarpus	*Sisyrinchium* spp.
Liatris scariosa	*Ceanothus pubescens*

A table similar to Table 7 was constructed for a group of 67 of the most important species of lowland forbs two of which are illustrated in Figure 35. The great importance of bedstraw (*Galium tinctorium*), for example, is shown by its occurrence as a society of the first class in 36 percent of the prairies and of the second rank in 16 percent (Table 8).

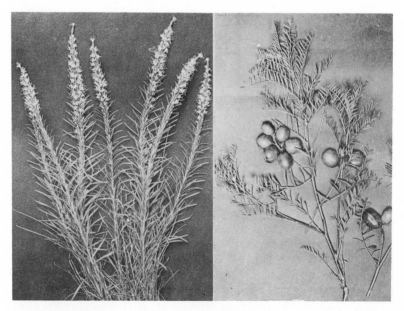

Fɪɢ. 34.—Blazing star (*Liatris punctata* [left]) and a branch of ground plum (*Astragalus crassicarpus*).

TABLE 8*

DISTRIBUTION AND RELATIVE IMPORTANCE OF THE CHIEF FORBS OF LOWLAND PRAIRIES
AND THEIR OCCURRENCE IN UPLANDS

Species	Low Prairie						High Prairie					
	1	2	3	4	5	0	1	2	3	4	5	0
Galium tinctorium	36	16	18	4	0	26	0	0	2	1	3	94
Fragaria virginiana	30	17	5	20	0	28	5	7	15	12	6	55
Steironema ciliatum	20	25	29	9	17	0	0	0	1	2	1	96
Aster salicifolius	27	17	25	12	10	9	0	0	1	2	0	97
Anemone canadensis	27	17	6	11	2	37	0	0	4	2	0	94
Solidago altissima	28	15	29	14	0	14	2	3	4	9	9	73
Silphium laciniatum	22	18	11	4	11	34	12	3	10	10	10	55

*The numbers under columns *1, 2, 3*, etc., give the percentage occurrences as societies of the first, second, etc., class, and the numbers under column *0* give the percent of prairies in which the species did not occur.

FIG. 35.—Two common plants of lowland: cutleaf rosinweed or compassplant (*Silphium laciniatum* [left]) and Canada anemone (*Anemone canadensis*).

Stiff marsh bedstraw (*Galium tinctorium*), a densely aggregated perennial with erect stems, is the most important. Attaining a height of 8 to 18 inches, this very tolerant species forms a dense understory in wet grasslands and in many big bluestem prairies. It occurred in 74 percent of the lowlands and ranked as a society of the first or second class in 52 percent. All of the other six species are abundant. All occurred in almost two-thirds of the low prairies, and all ranked as societies of the first or second class in 40 percent or more.

The next high-ranking species, in order, are:

Phlox pilosa	*Apocynum sibiricum*
Silphium integrifolium	*Viola papilionacea*
Helianthus grosseserratus	*Glycyrrhiza lepidota*
Liatris pycnostachya	*Pycnanthemum* spp.
Equisetum laevigatum	*Cicuta maculata*
Zizia aurea	*Leptandra virginica*
Teucrium canadense	*Asclepias verticillata*

Compositae furnished 34 percent of the leading forbs of the entire area; the Papilionaceae contributed 13 percent; and the Asclepiadaceae and Labiatae comprised 5 and 4 percent, respectively. More than 200 additional species occurred in less than 10 percent of the prairies, and some were found rarely, except in northern prairies. *Amorpha nana* and *Pulsatilla ludoviciana* are examples. Only southward did *Baptisia australis* and *Viorna fremontii* occur, and *Chrysopsis villosa* and *Sideranthus spinulosa* to the west.

Early descriptions, settlement, and studies in the eastern prairies of Ohio, Michigan, Indiana, Illinois, and Wisconsin compose a chapter in *North American Prairie* (Weaver, 1954). Studies in central and western prairies, from Minnesota to Texas, compose another chapter. A survey of the grasslands east of the Rocky Mountains, in Canada, and through Montana and the Dakotas to Texas and New Mexico occupies 8 of the 22 chapters in *Grasslands of the Great Plains* (Weaver and Albertson, 1956).

IV.

Ecological Studies in Prairie

The prairie offered a wonderful field for ecological studies. Much time has been spent in a study of the amount of underground materials, interception of rainfall, loss of water by runoff and of soil by erosion, and other phenomena.

UNDERGROUND MATERIALS IN GRASSLANDS

An excellent criterion of climate is the amount of plant material in the upper 4 inches of soil. It is here that stem bases occur, mostly in the upper 1 to 2 inches, where they arise from rhizomes or the crowns of plants and from a network of the attached roots. The amount of plant material is relatively small in the Chestnut soils of the plains of Colorado. It increases, both in weight and volume, in the better-watered Chernozem soils of south-central Nebraska and Kansas, and it is greatest in amount in the region of higher rainfall in the Brunizem soils of eastern Nebraska and southwestern Iowa (see Fig. 2).

Soils and vegetation are closely related, intimately mixed, and highly interdependent upon each other and upon the climate. The grassland vegetation has exerted a powerful influence as a soil-builder. Plants, which introduce the living, biological factor into soil formation, return much more to the soil than they take from it. In fact, their role in soil genesis is of such far-reaching effect that it is generally conceded that there could be no soil without vegetation. The great stability of the prairie, resulting in part from the long lifespan of many species, denotes a high degree of equilibrium between vegetation, soil, and climate.

A study was made by Shively and Weaver (1939) of underground plant materials at five groups of stations, centering—respectively—at Anita in southwestern Iowa, at Lincoln in southeastern Nebraska, at Nelson (90 miles southwest of Lincoln), at Phillipsburg in north-central Kansas, and at Burlington in eastern Colorado (see Fig. 36).

Mean annual precipitation decreased from east to west in the several areas as follows: from 33 to 29 inches, and then to 26 inches in true prairie, and from 23 to 17 inches in mixed prairie. Mean annual

63

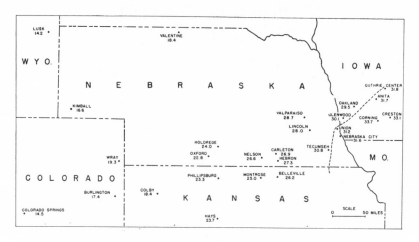

Fig. 36.—Map of portions of five states showing the stations where samples of underground plant parts were taken. The figures indicate mean annual precipitation. Stations east of the broken line of 31 inches of precipitation belong to the Anita group.

temperatures remained fairly constant, about 49° to 53° F. The prairies from which samples were taken, with very few exceptions (westward), were undisturbed except for annual mowing. The topography and soil of each prairie were typical of the surrounding region. The samples taken were 1 meter long, one-half meter wide, and 1 decimeter deep. All samples were taken from soils that were mapped as silt loam. The underground plant materials were secured by washing away the soil. Volume was determined by displacement of water and the oven-dried material was weighed. Series of samples were secured from nearly pure stands of little bluestem, big bluestem, mixed bluestems, blue grama and buffalo grass, and from certain other grasses (Fig. 37). Approximately 200 samples of prairie sod were collected from 24 stations, which were distributed over a region extending 600 miles east and west. More than 10 tons of soil were removed from prairies in Iowa, Nebraska, Kansas, and Colorado, and transported to Lincoln.

A series of 22 samples of little bluestem from the Anita, Lincoln, and Nelson areas yielded 3.15, 2.60, and 2.34 tons per acre, respectively, of underground plant parts. Underground yield for a series of 27 samples of big bluestem decreased from 4.54 tons per acre at Anita to 3.54 tons at Lincoln, and to 3.17 tons in the areas farther west. Yields of a series of 16 samples of mixed bluestems from the Anita, Lincoln, and Nelson areas were 3.42, 2.83, and 2.54 tons per acre, respectively. Data from each station in each area are recorded in the bulletin.

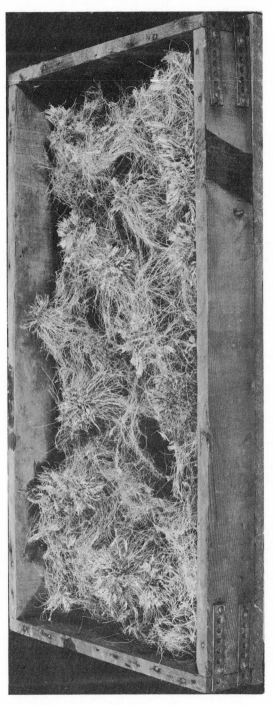

Fig. 37.—One-half square meter of underground parts of little bluestem (*Andropogon scoparius*), to a depth of 4 inches, at Nelson, Neb.

Statistical treatment of the data from each series showed that the correlation between volume and weight is very significant. The coefficients of correlation are 0.870, 0.874, and 0.699, respectively. The average volume of the underground parts of all upland grasses, if we use that of the Anita group as 100 percent, was 84 at Lincoln, 79 at Nelson, 73 at Phillipsburg, and 72 at Burlington. The percentage of dry weight, in the same sequence, was 100, 80, 73, 64, and 66 percent.

Correlation between the average dry weight and the mean annual precipitation at each group of stations also was found to be very significant. The coefficient of correlation for the little bluestem series was 0.642, 0.673 for the big bluestem series, and 0.827 for mixed prairie bluestems.

Blue grama and buffalo grass showed no consistent variations in yields from the Lincoln area, where they occurred sparingly (in poorer soils), to short-grass plains, where they are dominant. Weights of 82 samples of upland grasses showed a consistent decrease with decreasing precipitation westward. They were 3.35, 2.69, 2.43, and 2.19 tons per acre, respectively, the last figure being the average from the two driest stations.

Finally, the average percentage of organic matter decreased from east to west, from 7.14, 6.08, 5.41, and 4.42 percent to 2.67 percent. The mean from each area is significantly different from the mean of any other area. Thus some relationships between climate, vegetation, and soils were revealed.

RAINFALL INTERCEPTION

Much attention has been given to loss of water through runoff, evaporation from the soil, and transpiration, but water loss sustained through interception by herbaceous vegetation has received very little consideration. Interception of rainfall by prairie grasses, weeds, and certain crop plants was measured at Lincoln, Nebraska, during 1937 and 1938 (Clark, 1940), with the following abstract of results.

Such study is important for two major reasons. Plants, by preventing raindrops from striking the soil directly, have a marked effect upon decreasing runoff and erosion. Moreover, by holding a portion of the rainfall upon the surfaces of the leaves and stems until it evaporates, a considerable amount of water is prevented from reaching the soil, where part of it would eventually be available to the roots of the plants. Thus a very important loss to the vegetation results.

In order to determine the magnitude of this loss of water, numerous methods have been devised, and two methods have been found which readily lend themselves to field studies with prairie vegetation, certain crop plants, and weeds. A meter quadrat is marked out on the surface of the soil beneath the plants. In it are placed five pans, each 1 meter long, 4 centimeters wide, and 5 centimeters deep (Fig. 38). The surface covered by the pans represents one-fifth of the total surface of the quadrat. By means of conveniently spaced, permanent cross wires and a wire mesh in the bottom of each pan, it is possible to place cut plants in the pans in their normal position. When necessary for proper placing of the pans, plants are cut off at the soil surface and inserted in the pans in the same position that they previously occupied. Water is then applied by means of large bottles, equipped with sprinkler tops, and the amount is expressed as 1 inch per hour, one-half inch in 30 minutes, or in smaller amounts. Such factors as light, air temperature, humidity, and wind movement are measured during the progress of the experiment. The amount of water caught in the pans represents one-fifth of the water not held by the plants, and thus it is possible to express the amount of water intercepted in percentages of the total amount applied.

Fig. 38.—Interceptometers, partly withdrawn to show something of their construction. The grass is big bluestem.

Such mat-forming plants as prostrate pigweed, knotweed, etc., are cut off at the soil surface and placed in their natural position upon a quarter-inch mesh screen that is 1 square meter in area. The screen is then suspended over a large pan and water is sprinkled upon the plants at predetermined rates. The interception capacity of the plants is calculated as before. The effect of wind movement is readily shown by use of an electric fan. Thus the plants are under practically natural conditions. It is not claimed that all the factors which characterize a rainstorm are present, but the methods lend themselves to simulate natural conditions.

The amount of water intercepted by herbaceous plants is often surprisingly large. Wheat, when fully developed, was found to hold from 50 to nearly 80 percent of the water applied, depending upon the rate of application. With alfalfa (*Medicago sativa*), interception was as high as 89 percent during a light shower and as low as 26 percent during a heavy rain. An open growth of needlegrass (*Stipa spartea*) in an upland prairie intercepted approximately 50 percent of the water applied at the rate of one-fourth inch in 30 minutes. Prairie dropseed (*Sporobolus heterolepis*) gave somewhat similar results, but little bluestem (*Andropogon scoparius*) intercepted from 50 to 60 percent of the water applied at the rate of one-half inch in 30 minutes. In low prairie, composed chiefly of big bluestem (*A. gerardi*) and switchgrass (*Panicum virgatum*), with flower stalks fully developed, the interception at different rates of application was: 1 inch in an hour, 47 percent; one-half inch in 30 minutes, 57 percent; one-fourth inch, 67 percent; and one-eighth inch, 81 percent for 30-minute periods (see Fig. 39). Bindweed (*Convolvulus arvense*) intercepted 17 percent of water applied at the rate of one-half inch in 30 minutes, 30 percent when one-fourth inch was applied, and 50 percent when one-eighth inch was used. For buffalo grass (*Buchloe dactyloides*) the results were: one-half inch in 30 minutes, 31 percent; one-fourth inch, 46 percent; and one-eighth inch, 74 percent. In all the experiments it was found that wind, through its influence upon evaporation, has a marked effect upon the percentage of interception.

Results show that the amount of water held upon the surfaces of leaf and stem and prevented from reaching the soil is very great. They also show that the amount of water thus held depends largely upon the rate at which the water falls, and—to a certain extent—upon the environmental conditions, especially wind movement. In the plants studied, little water ran down the stems and thus reached the soil. So far as use by the vegetation is concerned, the water intercepted represents a loss, which over large areas becomes enormous. For example, when a thick growth of bindweed intercepts 13 percent of one-half inch

Fig. 39.—Interception of water by the foliage of big bluestem.

of water in 30 minutes, the amount withheld by the plants from reaching the soil is 7.5 tons per acre. Wheat, in intercepting 52 percent of a similar rainfall, causes a loss of over 29 tons of water. When an inch of water falls during an hour, buffalo grass intercepts over 28 tons per acre, and prairie that is composed chiefly of big bluestem may intercept as much as 53 tons per acre.

Water is held upon plants in the form of thin films or as drops which form on the surface, at the tips, or along the margins of the leaves.

Water also adheres to the stems. The extent of the leaf surface and the number of levels at which water may be held are important factors in determining the percentage of interception. Prairie vegetation has a foliage surface that is 3 to 20 times as great as the soil surface beneath it; leaves are displayed at many different levels (Clark, 1937).

RUNOFF AND EROSION

The same cover of vegetation that intercepts the rainfall exerts a profound effect upon the force with which the raindrops strike the soil and upon their entrance into the soil by absorption. In prairie where there is a cover of grass, the force of the rain is broken by the foliage of the grass and other herbs and by the litter of fallen leaves and stems beneath. Hence the rain does not beat directly upon the soil. The lodgement of the undecayed materials among the stems of the grasses forms an intricate series of minute dams and terraces which tend to hold the water until it can percolate into the soil. Abundant humus creates a spongelike condition in the topsoil, and this increases its capacity to absorb and hold water. Hence runoff in the prairie is usually slight unless the rains are heavy. Even during a heavy rainfall, the water that runs off is usually clear since the soil is firmly held in place by the bases of the plants, by their widely spreading and much-branched rhizomes, and by their widely and deeply spreading root systems—as may be illustrated by a single example. During a rainfall of 5 inches over a period of two days, the runoff from a native prairie on a 5-degree slope was only 3 percent, all clear water. On a similar slope, only 35 feet distant but in wheat stubble, the runoff was 28 percent, and more than 1/100 inch of surface soil was washed away (Weaver and Noll, 1935).

It is only when the vegetation is closely grazed and the amount of roots and rhizomes diminished that serious erosion begins. Other factors being equal, the intensity of erosion is directly proportional to the decrease in the amount of vegetation, both above and below ground. Upon prairie areas from which the vegetation has been largely or completely removed by overgrazing, the raindrops beat upon the bare soil like millions of little hammers. The soil, already trampled by grazing animals, is further compacted and its absorbing capacity is further reduced. The cohesive force between the soil particles is lessened as the surface becomes muddy. The particles shift their position under the effect of the beating rainfall and fill up the soil pores; thus the absorbing capacity is reduced. The excess water accumulates on the surface and, on running off, removes with it the surface soil particles, the humus with its micro-population, and the dissolved salts.

Grassland sod is a great conserver of rainfall; the amount of runoff water is relatively small and the soil is firmly held against the forces of erosion. In experiments by Weaver and Harmon (1935), areas of prairie sod 1 meter long and one-half meter wide were secured intact in frames to a depth of 4 inches. They were placed at an angle of 10 degrees and the soil was washed away under controlled conditions. Big bluestem in Wabash silt loam resisted erosion longest, but bluegrass in similar soil resisted erosion almost equally well. Needlegrass was somewhat less effective than little bluestem in holding the soil.

The rhizome-root network of the native grasses was retained almost in place after the soil had been removed, and that of bluegrass sank to the bottom of the frame. The holding power of bluegrass was clearly less after the first 3 inches of soil were removed.

Extensive experiments have been made by Kramer and Weaver (1936) on the relative efficiency of roots and tops of plants acting together, and underground parts alone, in preventing soil erosion. Numerous crops of cultivated field and garden, pasture crops, weeds, and native grasses were used.

A method of securing samples of undisturbed field soils with crops uninjured and in all stages of development has been devised. The samples were 1 meter long, 0.5 meter wide, and 1 decimeter deep, and weighed 170 to 200 pounds (see Fig. 40). Samples were taken in pairs and transported to a washing rack with a slope of 10 degrees. Tops were removed from one sample, after being clipped close to the soil surface, but were left intact in the other. The time required to erode the soil of the two samples under the same conditions of watering was ascertained. Water was applied at the rate of 12.7 gallons per minute, under a total force of 1 pound, over an area of approximately 1.5 square inches; the hose was moved constantly and uniformly.

Bare soil, consolidated by occasional watering during four weeks, eroded in 16 to 18 minutes. In soil held by roots of seedling sunflowers, millet, or Sudan grass that were 30 days old, the erosion time increased to 2 to 4 hours. With tops also intact, 35 percent or more of the soil was left after 9 to 12 hours of watering.

Lowland soil that was planted to winter wheat eroded in late fall in 7 and 11 minutes, without and with tops, respectively. With the growth of the crop, the underground parts alone more than doubled the efficiency of the soil in resisting erosion. Protection afforded by maturing tops increased resistance 9 to 10 times (Fig. 41). The early top-root ratio of 1 : 1.6 gradually increased to 1 : 7, but fell after harvest to 1 : 1.2. Thus the great importance of plant cover in protecting the soil was clearly revealed. A good stand of Sudan grass was more efficient

Fig. 40.—Sample of winter wheat, taken on Nov. 10. The plants are about two inches tall and much bare soil is exposed.

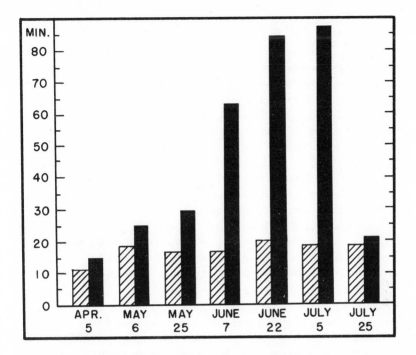

Fig. 41.—Relative efficiency of underground parts of winter wheat without tops (hatched) and with tops (black) in holding the soil against erosion. The crop was harvested before July 25.

in retarding erosion than any other field crop examined. This resulted from the rapid growth of an abundance of long leaves, tillering—which promoted close spacing of stems—and early development of a strong, fibrous root system. Erosion time without tops was increased, when plants were mature, more than 12 times.

Well-established Hungarian bromegrass was the most efficient soil protector found among pasture plants. The root mass approached native prairie grasses in its efficiency as a soil binder. Rye drilled thickly for fall and spring pasture is an efficient cover crop during a season when most crops do not thrive.

Prevention of erosion does not depend so much upon vertical thickness of cover as upon one that is widely spread and continuous. A single leaf on the soil is very effective in protecting the soil directly beneath it. Sorgo retarded erosion well because of the great production of long leaves which broke the force of the water and conducted it gently to the soil. Strong stem bases and abundant coarse roots anchored the soil. Erosion time with tops was 98 minutes.

The character of the crop is a principal factor in erosion control. The effect with the plant cover intact exceeds that of underground parts alone 3 to 7 times. Maximum protection among crops was afforded by winter wheat and sorgo; oats and alfalfa offered less protection. The most formidable line of defense by the grasses against erosion is above ground, although the soil also is held in a remarkable manner by roots and rhizomes. The foliage of prairie cordgrass is especially adapted to protect the soil, even against rushing flood water.

Soil erosion is a national menace. It is a complex problem in which the pertinent factors of climate, slope, soil, and vegetation must each be analyzed and evaluated for different areas and regions. A full understanding of the scientific principles underlying the process will be had only after long-continued research.

Erosion is one of the interactions between climate, vegetation, and soil, in which the plant cover is a decisive factor. There was no problem of accelerated soil erosion in the West until much of the grassland was broken for cropping or was weakened by continuous overgrazing.

Soil is as much a product of vegetation as vegetation is a product of soil. Throughout the centuries vegetation has favorably influenced the development of soil. Its presence in the soil, whether alive or dead, profoundly affects soil structure, water absorption, percolation, and water retention, while the mantle of grassland above the ground protects the earth like a garment.

Runoff and erosion have been measured from entire watersheds, and, more recently, by the runoff-plot method at Lincoln during 1934 and 1935. They were measured on prairie, pasture, and cultivated land. Enclosed plots 3 feet wide and 33.3 feet long were used in the following studies. Natural rainfall was supplemented by watering (see Fig. 42).

Runoff on a 10-degree slope from 26.9 inches of rainfall during fifteen months (1934 and 1935) was 2.5 percent from prairie, 9.1 percent from overgrazed pasture, and 15.1 percent from a pasture entirely bared by close grazing. The soil was a silt loam and the erosion plots, resulting from special previous fencing, were only a few yards distant. No measurable amount of soil eroded from the prairie, and only a small amount from the pasture, but 5.08 tons per acre were lost from the bare area. Both years were years of drought, and consequently runoff and erosion were light.

Numerous experiments have shown that where there is a good cover of grass there is no serious problem of erosion. But where the cover of grass is broken or removed by excessive grazing, erosion is the inevitable sequel. Nature, if unhindered, will repair the cover if soil erosion has not progressed too far, but once the good topsoil is washed away, restoration of former conditions requires very long periods of time. Pasture improvement is a chief weapon against erosion.

Fig. 42.—View of prairie in a runoff plot on a 5-degree slope on May 2, 1935 (above). Alfalfa 5 inches tall in a runoff plot on the same slope on May 2, 1935.

Runoff on a 5-degree slope from 12.9 inches of rainfall during a period of eleven months was 1 percent from prairie, 12.1 percent from wheat field, and 17.8 percent from fallow land. The soil was a silt loam. No measurable erosion occurred in the prairie, but 0.52 ton of soil per acre eroded from the wheat field and 2.6 tons from the fallow land. In another experiment, 5 inches of water were applied to prairie and wheat stubble, and 4 inches to fallow land, during a period of two days. Runoff was 3.1, 27.6, and 23.2 percent, respectively, and soil erosion was insignificant: 1.29 and 1.75 tons per acre.

Other runoff plots with a 5-degree slope were tested in prairie and in alfalfa. Four inches of water, which were applied to each plot in spring (when the alfalfa was 5 inches tall), resulted in 5.9 percent runoff in prairie and 40.8 percent in alfalfa. No erosion occurred in the grassland but 0.72 ton of soil per acre was removed from the field.

Runoff resulting from the application of 3 inches of water in 1.5 hours on May 10 on a 7-degree slope on Lancaster sandy loam was nil from burned prairie, but 20 percent from broken prairie that had been cropped to corn for a period of six years. Topsoil lost by erosion was 12.2 tons per acre (Weaver and Noll, 1935).

A soil covered with its natural mantle of climax vegetation represents conditions that are most favorable to maximum absorption of rainfall and maximum erosion control. Soils that have been depleted of their organic matter and are poor in structure are less absorptive and are easily eroded. Methods of increasing the use of grass and other thickly growing crops that furnish a cover similar to the prairie should be intensively studied. Thus more of the rain may be retained where it falls and more of the soil on the slopes may be held in place. Experiments at the Federal Erosion Experiment Stations showed that, under many conditions, erosion can be reduced enormously or almost completely controlled with adaptable measures that involve a cover of vegetation. One of the chief essentials of erosion control is the increased use of grasses.

THE WONDERFUL PRAIRIE SOD

The prairie sod had been penetrated many times and in various places, but a study of its horizontal spread was finally investigated in 1961/1962. The following account and figures are taken from the *Journal of Range Management* (Weaver, 1963).

Few studies have been made upon the rhizomes or other plant parts in prairie sod. A notable exception is the excellent study by Mueller (1941), who traced the development of the rhizomes of several prairie grasses and forbs from seedlings to the adult stage, and thus through a complete annual cycle. Rate of vegetative spread also was ascertained.

The advent of the great drought of 1933 to 1940 offered an exceptional opportunity to ascertain the responses of native plants to extremely adverse conditions. The role played by rhizomes of grasses in endurance of drought—and frequently their recovery from it—was very impressive. Indeed, only then did the great importance of the part of the prairie in the upper 4 inches of soil—the prairie sod—become clear to me.

Plant parts in prairie sod are protected from sudden and extreme changes in temperature. They are scarcely harmed by frost or severe cold of winter, driving hail, tornadoes, or prairie fires. They endure ravages by grasshoppers and greatly prolonged drought. To prairie sod, only the plow or long-continued close grazing are lethal.

The development of an extensive system of rhizomes not only enables grasses to spread widely into open spaces, but also to invade underground and place new growing points between or beneath other

Fig. 43.—Surface view of a square foot of rhizomes of big bluestem after the soil was washed away and the roots removed.

plants, which then often are replaced. Moreover, rhizomes afford an excellent place for food accumulation, and some retain life even after above-ground parts have long since disappeared and after root systems have been greatly weakened or have died.

Translocation and storage of reserves in the underground organs of grasses has been well summarized:

> Certain carbohydrates (mainly sugars, fructosans, dextrins and starch) have been shown to be the principal reserve substances in grasses. These materials are elaborated by the leaves in excess after flowering, and are subsequently translocated to the roots and rhizomes, where they are stored to be drawn in the following spring for the production of new top growth. Nitrogen and mineral elements (though not being reserves in the true sense of the word but merely nutrients) are likewise translocated in autumn from the aerial parts to the underground system where they are stored over winter (Weinmann, 1948).

Big bluestem produces great numbers of branched rhizomes that are 3 to 9 millimeters in diameter and compacted into dense mats about 2 inches thick at depths between 1 and 3 inches (see Fig. 43). Their length per square foot of sod averaged 55 feet. Tensile strength was 55 to 64 pounds for individual rhizomes. Indian grass is similar to the preceding grass in total length of rhizomes and in their position in the sod, but it is less efficient in spreading.

Switchgrass has rhizomes 3 to 7 millimeters thick and 1 to 2 feet long. It differs from big bluestem by forming an open framework 2 to 5 inches deep, but sometimes 8 inches deep (see Fig. 44). A square foot of soil contains about 50 feet of much-branched and interwoven rhizomes. Single rhizomes exhibited a tensile strength of 80 to 132 pounds.

Beneath mature plants of prairie cordgrass, the soil to a depth of 6 to 10 inches contains a mat of coarse, woody, much-branched rhizomes 4 to 8 millimeters thick. Lateral branches originate at various depths from vertically placed rhizomes. Some exceed 2 feet in length, and lateral spreading is rapid. Compared with big bluestem, the rhizomes are much coarser and the network much more open, but the latter are greater in the vertical dimension. Total length of rhizomes averaged 80 feet per square foot. On moist lowland, rhizomes are common on many grasses, sedges, and rushes.

Of the rhizome networks of the four chief lowland grasses, that of big bluestem is the shallowest. That of Indian grass is very similar in the vertical dimension. Switchgrass forms an open network with a depth of 3 to 8 inches. Prairie cordgrass has a framework of variable thickness, ranging from 2 to 8 inches.

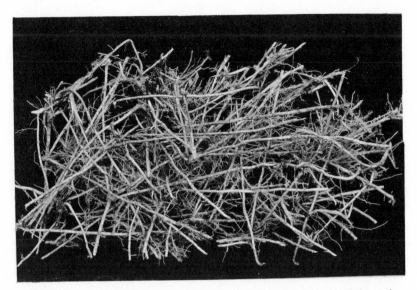

Fig. 44.—Coarse rhizomes on the bottom of an 8-inch-thick mat of the underground framework of switchgrass on a flood plain near Hamburg, Ia. Under a single square foot of soil there were 50 feet of rhizomes.

Rhizomes furnish abundant room for storage near the food factories above, and they are close to the water and nutrient supplies around and beneath them. Fifty feet of rhizomes with an average diameter of 4 millimeters have a volume of 11.7 cubic inches. Above these food-storage reservoirs the tall, coarse grasses develop shoots rapidly, followed by an enormous expanse of leaves.

Dominants of uplands are mostly, but not all, bunch grasses. The most abundant of these is little bluestem. Its compact bases are closely compressed in the surface 1.5 to 2 inches of soil. Roots begin at the bases of buried stems and spread outward in all directions, thus binding the stems together and firmly anchoring the bunch (Fig. 45). Roots

Fig. 45.—Plot of little bluestem prairie, about 6 square feet, that shows stem bases and roots in the surface 4 inches of soil.

spread laterally in great numbers, to at least 2 feet in the surface 4 inches of soil. Numerous other plants, especially other grasses, grow between the large bunches and also occupy a part of the upper soil. Their roots and rhizomes furnish a considerable part of the 2.6 to 3.15 tons of plant materials per acre in the upper 4 inches of the bluestem sod. This is essentially true also of bunches of prairie dropseed and needlegrass, which also maintain upland communities.

Seasonal Aspects and Stability

The prairie presents four seasonal aspects: prevernal, vernal, estival, and autumnal. Each aspect is characterized in part by the blossoming of a different group of species. A fifth aspect hiemal, occurs in winter, and extends from late October until early March. It is characterized by the dried grasses and forbs, many of which remain intact for a long time. As a result of natural deterioration, and augmented by the wind and assisted by the weight of ice and snow, the once-standing vegetation gradually returns to the surface of mother earth. Here it forms a protecting blanket over the living parts within and beneath the surface of the soil. Before the coming of the white man, the accumulated debris was removed by the occurrence of fires, resulting from lightning or set by Indians, which burned over vast areas. Accumulation of natural mulch year after year greatly affects the composition of the vegetation (Weaver and Rowland, 1952).

Species of the several aspects in prairie have been listed by Thornber (1901) and Steiger (1930). A full description of each aspect has been given by Weaver and Fitzpatrick in *The Prairie* (1934), and also by Weaver in *North American Prairie* (1954a). Both accounts are beautifully illustrated.

About a dozen very early blooming forbs characterize the prevernal aspect, but the vernal aspect presents about 40 species. The estival or summer aspect has the largest number, about 70 in all, while species blooming in the autumn are about 40 in number. Each seasonal aspect also is characterized by the degree of development of the grasses. Even the winter aspect does not lack charm and is often one of splendor.

Stability denotes a high degree of equilibrium between the vegetation and its habitat under the control of the existing climate. It does not preclude minor changes in the abundance of the constituent species. This regularly occurs from year to year as a response to the irregular variations in the factors of the habitat-complex. The phenomenon, however, is often more apparent than real. The more important species are long-lived and continuously present. They may be either less or more conspicuous because of their lack of abundance of flowering and fruiting, depending upon locally unfavorable or favorable conditions. Conspicuous fluctuations occur mostly among annuals or other relatively short-lived plants. Although the details of the pattern of the prairie mosaic may change, the shiftings are of minor importance. The relative constancy of the number of plants over a long period of time and the ordinary fluctuations within relatively narrow limits indicate the high degree of balance or stabilization. There are no great waves of emigrations, neither are there immigrations; the prairie is a closed community, and invaders—with rare exceptions—are excluded.

Large tracts of prairie are almost uninvaded by weeds except to the extent that trails or roads have been made through them or because soil has been washed into the ravines from adjacent fields. It is indeed impressive to find these relict areas of prairie entirely uninvaded, although surrounded on all sides by cultivated crops with their accompanying annual weeds or by pastures with their usually longer-lived weedy flora. They are free from invasions, although the kinds of invaders are numerous and their methods of competition diverse. In fact, the number of possible invaders is quite as large as the number of prairie species themselves. A list of the immigrant flora of Iowa alone contains 263 species (Cratty, 1929), and Shimek (1931) stated that 265 species make up the bulk of the prairie flora of Iowa. Steiger (1930) found 237 species of prairie plants on a single section (640 acres) of land near Lincoln, Nebraska.

Stability is increased by the long span of life of many prairie species. Perhaps only about 5 percent are annuals. The dominant grasses, once established, retain their vitality for many years. The climax vegetation is the outcome of thousands of years of sorting out of species and adptations to the soil and climate. In fact, it is more than this, for the vegetation itself has had no small part in determining the physical, chemical, and biological properties of the soil. It also has reacted upon the climate. The vegetation represents not only the present edaphic and atmospheric factors but also those of the past. Climax prairie is in close adjustment with its environment.

Unless disturbed by man, and barring the entrance of a new dominant from another region, the prairie will maintain possession until there is a fundamental change in climate or until a new flora develops as the outcome of long-continued evolution (Weaver and Flory, 1934).

V.

Underground Activities in Prairie

Earlier students of ecology had little information on what portion of plants occurred underground and even less about the activities of plants within the soil. Consequently, several years of research by the writer and his graduate students were spent on a study of these problems (Weaver, Kramer, and Reed, 1924).

Development of Seedlings, Tillers, and Rhizomes

Grasslands are relatively dry lands; drought, at least in the surface soil, is always imminent. The necessity for a seedling to make immediate and extensive contact with deeper, moister soil is apparent. The water relationship usually is controlling, and the success or failure of the seedling depends largely upon its ability to develop an absorbing system that is adequate in extent and activity to furnish the necessary water supply. Underground plant parts play a role of major importance in the starting of populations. Rapid germination and elongation of the primary absorbing system are requisite to successful establishment. The production of tillers is usually successful only when the secondary root system penetrates into the moist soil and meets the increased demands for water (Weaver, 1930).

Climax grasslands, whether of the dense, tall-grass type or of the open, desert-plains type, are communities in which the least-abundant necessary factor, water, is utilized in its entirety. There may be an excess for long or short intervals, but the constant struggle for an adequate supply limits the population in numbers and dwarfs the individuals in stature. That even the dominant grasses are seriously handicapped by this shortage of water is readily shown by removing the vegetation from all but the central portion of a square-meter quadrat. The remaining plants benefit greatly by the extra water supply thus afforded, and their growth in terms of dry weight is often more than doubled.

The bunch and sod habits, resulting from tillering and rhizome production, increase the plant cover even when conditions are unfavorable for the establishment of seedlings. Tillering in most species of grasses begins simultaneously with the development of the secondary root system, and usually three to six weeks after germination. The period of tillering before new roots become well established in the moist layers of soil is a very critical one for the plant. The transpiring area has developed somewhat ahead of the root system that is in contact with the deeper soil. Next to germination, it is the most critical time in the life of the plant. But once the secondary root system is established, the chances for the completion of ecesis are fairly certain.

The degree of tillering is a fair index to the rapidity of growth and the extent of the root system. Crowding delays or prohibits tillering. Some grasses, such as Indian grass and switchgrass, when grown in dense cultures, tiller with difficulty and may develop only single stalks even after a long period for growth.

In many species, possession of space is obtained by supplementing early tillering with the production of rhizomes. Western wheatgrass, for example, occupies only the soil beneath the plant, but rhizomes, with their new roots, appear at an early age and the sphere of its activities is rapidly extended. The rhizome is a valuable asset in permitting the species to consolidate its gains. In both types of vegetative propagation, better and more prolonged provision for the offspring is assured. The parent furnishes water and nutrients until the new individual is established, often in the presence of competitors.

The increased vigor of the plant during successive years is clearly related to better development of the parts underground, and especially to the reserve supply of food accumulated in them. Among dicotyledonous species, where roots and rhizomes are large and the storage capacity is thus greatly increased, the rate of development is marked. In our experimental gardens, prairie sunflower (*Helianthus laetiflorus*), for example, reached a height of 1 inch early in June of the first year; by the second year the plants were 12 inches tall; and on the same date in the third year, the plants were 20 inches tall. *Liatris punctata*, blazing star, developed a deep taproot the first summer that was out of all proportion to its shoot. In August, plants only 5 inches tall, and with only 2 leaves, possessed taproots 33 to 38 inches deep which had accumulated some reserve food. It developed slowly through the years but it is very long-lived (Clements and Weaver, 1924; Blake, 1935).

EXTENT AND LONGEVITY OF SEMINAL ROOTS

Grasses possess two distinct root systems. The primary or seminal root system begins development immediately upon the germination of the seed and consists of one to several main roots and their branches; the number, varying with the species, is often only one to three.

The young plant is entirely dependent upon this primary root system for water and soil nutrients. Curiously, the seminal roots were often designated as temporary roots, even in special books on grasses, and they were believed to exist for only a short time. Studies of seminal roots of perennial grasses were few, however, and usually any mention of their development was only incidental.

Grasses were grown from seed, in appropriate wooden boxes that were lined with galvanized iron (with one side removable), and the boxes were filled with loam potting soil. The development of both tops and roots of 14 species of native and introduced perennial grasses, such as the bluestems, needlegrass, and species of wheatgrass, was examined at four stages of growth.

In all species the seminal root system extended to 10 inches in depth during the first 21 days after planting, and in some species to 27 inches after 41 days. Depth had increased but little at 59 to 74 days, but almost all of the seminal roots were functioning at a time when the abundant tillers were one-half to two-thirds—or even fully—as tall as the parent culm. One to four nodal roots of the secondary root system developed on the parent culm of nearly all of the grasses before tillers appeared.

Seminal roots were usually very fine and most of the main seminal roots were 0.3 millimeter or less in diameter. Nearly all were profusely branched. Little or no decortication of the main root had occurred on most species after 21 days, and, except for one species, the cortex of the branches was intact. After 53 to 74 days, when the plants were 5 to 16 inches tall, there were many deeply penetrating nodal roots and numerous tillers one-half to two-thirds the height of the parent culm.

At 90 to 123 days, the grasses were 9 to 19 inches tall, and the nodal roots of two species had blossomed. Despite 6 to 60 nodal roots, the seminal roots, which were marked by aluminum bands, extended deeply. Microscopic examination indicated that only about one-third of the seminal root system appeared normal in one small group, but on the remaining species one-half to two-thirds of the root system remained intact. Maximum penetration of seminal roots, 2 to 3 feet, was attained by some species.

Howard (1924) stated that on the black soils of India the upper soil dries so rapidly after sowing that there is hardly any development of the secondary root system and practically no tillering. The crop develops slowly on the primary roots. Other researches support this view. In our experiments, normally developed plants of several species grew 77 days and to heights of 8.5 to 14 inches, on the seminal roots alone, when the nodal roots were repeatedly excised and not permitted to develop (see Figs. 46, 47). One plant of buffalo grass developed 20

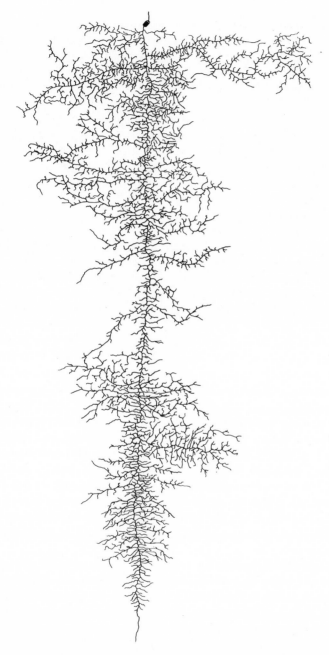

Fig. 46.—Seminal root system of *Andropogon gerardi* 70 days after the seed was planted. It is 21 inches long.

Fig. 47.—Top of *Andropogon gerardi* that was grown on the root shown in Figure 46. It is 8 inches tall, and all but 2 of the leaves are green and thriving. Note the single seminal root and band.

square inches of leaf surface in 60 days on a single, hairlike seminal root. The three seminal roots of wheat, with nodal roots excised weekly, were—after the period of blossoming—3 to 4 feet deep. This was 10 to 23 inches deeper than those of the control plants. The roots also rebranched more than those of the controls.

Hence, we must conclude that seminal roots of perennial grasses are not temporary. Those of 14 perennials remained alive and active 3.5 to 4 months; and possibly under natural field conditions, where growth of seedlings is much slower, they would remain so at least throughout the first season of growth (Weaver and Zink, 1945).

Annual Increase of Roots and Tops

How much root material is produced annually? What is its relationship to the growth of tops? How many growing seasons are necessary for maximum production? These are fundamental questions in a study of grasslands. To answer these enquiries, separate lots of range grasses were employed. They consisted of a representative tall grass (*Andropogon gerardi*), a mid grass (*A. scoparius*), and a short grass (*Bouteloua gracilis*). Three lots of each grass were grown from seedlings that were transplanted without disturbing the soil, into large steel drums that were filled with a well-screened potting soil at optimum water content. For blue grama, a lighter soil was provided by adding one-third the volume of sand to the potting soil. Thus, even during the third year of growth, all grass roots were kept in the containers and roots of other plants were kept out.

The drums were 34 inches high and 22.5 inches in diameter, and each had a 58-gallon capacity. They were placed in a trench and the plants were grown in a bluegrass lawn. Adequate provision was made

Fig. 48.—Root systems of *Bouteloua gracilis* on September 5 of the second summer. Root depth is about 32 inches.

for drainage. Nine drums were used, three for each species. The contents of five flower pots, each with 10 seedlings of big bluestem, were transferred to each of three drums without disturbing the roots. A similar lot, with seedlings of little bluestem, was planted in three other drums. The three drums for blue grama received twice this number of seedlings. The plants were properly watered, the surface soil was mulched, and all of the seedlings flourished. At no time did they show wilting.

At the end of each growing season the root systems from one drum of each species of plant were washed free of soil and photographed, and the oven-dry weight was obtained (see Fig. 48). Big bluestem produced 152 grams of roots and 306 grams of tops the first summer. It reached approximately its maximum root development the third summer. Maximum production of both little bluestem and blue grama was attained in two growing seasons, as is shown in Table 9.

TABLE 9

DRY WEIGHT OF TOPS AND ROOTS OF THREE SPECIES OF GRASS DURING THREE SUMMERS' GROWTH, IN GRAMS

Species	Dry Weight of Tops			Dry Weight of Roots		
	1943	1944	1945	1943	1944	1945
Andropogon gerardi	306	518	804	152	261	321
Andropogon scoparius . . .	388	790	906	89	166	160
Bouteloua gracilis	283	394	441	76	118	92

The underground materials produced in three years were very similar in amount to those ascertained earlier in typical mature stands of native prairie. Roots alone yielded approximately 5.5 tons per acre after three years' growth of big bluestem, 2.7 tons in little bluestem, and 1.6 tons in blue grama. In this experiment, 78, 69, and 80 percent of the root systems of the three grasses, in the preceding order, occurred in the surface foot of soil. This is in accord with later field studies.

Other topics discussed in this report were the effects of the roots of grasses in producing a granular soil structure, the time required for the completion of this structure-forming effect, the decrease or loss of soil granulation under continuous cropping to cereals, and the restoration of good structure and fertility by growing perennial grasses (Weaver and Zink, 1946).

LENGTH OF LIFE OF ROOTS

In our midwestern climate, the freezing of the soil each fall witnesses the death of the above-ground vegetation; the prairie retreats under-

ground. Year after year new shoots replace the old ones in this vegetation of long-lived perennials. There was almost no information in 1945,
however, on what portion of the root system is maintained and over
what period of time. From studies made in the eastern United States, it
was concluded that at least half of the root system of certain grasses, such
as bluegrass, is newly generated each spring (Sprague, 1933). Stuckey
(1941), who examined the roots of ten species of cultivated grasses, stated that for some species the whole root system is regenerated annually.

In our studies, seeds of ten species of perennial range and pasture
grasses were planted in triplicate lots in containers that were large
enough for ample root development. A removable extension at the top
of each container was filled with sandy loam soil that was easily washed
away when the extension was removed, thus exposing the roots for
examination (see Fig. 49). Small bands of aluminum or of very thin,
pliable sheet tin were placed around individual roots of the secondary
system. Preliminary banding had been done in 1929 and the methods of
procedure outlined (these were successfully used by Stoddart in 1935).
The first banding by Weaver and Zink (1946a) was done between
April 26 and May 14, and a second banding, involving new roots of the
same plants that developed later, was done between June 7 and June
15, 1943. By keeping the roots moist with a spray of water and then
covering them with dry soil, which was immediately watered, they were
unharmed. A total of 3,424 roots of 181 plants were banded to ascertain their longevity. Some were examined at the end of the first and
second years, and the remainder were examined at the end of the
third growing season. (The bands were still bright and unrusted.)

Survival of the banded roots of Hungarian bromegrass (*Bromus
inermis*) at the three yearly examinations was found to be 92, 84, and 36
percent. Survival of switchgrass (*Panicum virgatum*) and western wheatgrass (*Agropyron smithii*) after the second summer was 100 and 42 percent,
respectively. Ninety-seven percent of the banded roots of crested
wheatgrass (*Agropyron cristatum*) survived the first summer and 75 percent survived the first year. After three growing seasons, 81 percent of
the roots of big bluestem (*Andropogon gerardi*) were intact, but none of
the less stable, nodding wild-rye (*Elymus canadensis*). Field observations
over a long period of years indicate that this wild-rye is a species that
depends for its permanency of occupation more upon rapid reproduction from seed than upon the length of life of the individual.

Losses in all species were gradual, and after three growing seasons
the survival of banded roots was as follows: blue grama (*Bouteloua
gracilis*), 45 percent; side-oats grama (*B. curtipendula*), 14 percent; and
little bluestem (*Andropogon scoparius*) and needlegrass (*Stipa spartea*),

Fig. 49.—Roots of *Panicum virgatum*, about 3 months old, exposed for banding.

each 10 percent. The average number of roots produced by individual plants varied from 175 to 882 at the end of the third summer.

Compared with the total number of roots, losses among the banded roots were small to negligible. They often amounted to only 2 to 8 percent of the total number of living roots. Certainly, the loss of a few score roots among hundreds of others would have little effect upon a plant. The writers have experimentally determined that, under usual conditions, the removal of half of the root system had little harmful effect upon the growth of several species of grasses. Whether or not the roots that develop in the second or third year are of longer lifespan than those of the first remains to be ascertained, but this seems probable. Nearly all of the species of grasses studied have a long life and show great permanency of occupation, two features that add greatly to the stability of the prairie sod.

Depths at Which Plants Absorb Water and Nutrients

The great extent of the root systems of most plants that had been examined and their usual thorough occupancy of the subsoil led to investigations concerning their activities in subsoil and parent materials. Although the capillary movement of water in soils had by this time been shown to be much less than formerly was supposed, experiments were devised to check all movements of either water or nutrients from one soil layer into another, except through the roots. The soil—after being brought to the desired water content—was firmly compacted in the experimental containers and then sealed with a layer of wax. This consisted of 85 percent paraffin or parowax and 15 percent petrolatum. It was applied hot, so that it penetrated a little into the soil, and when it cooled the wax clung tenaciously to the soil particles. The seal was only 2 to 3 millimeters thick. When it had cooled and hardened, another layer of soil, usually 6 to 12 inches thick, was added and the process was repeated until the container was filled. The roots of various native and crop plants grown in these soils were distributed evenly throughout the several hermetically sealed layers, and penetrated the wax seals without difficulty. The seal had no apparent effect on root development.

This method furnishes an immediate means for determining the amount of water or nutrients removed at any given level to which the roots penetrated. Moreover, by using a series of containers in which plants of the same age were grown, it was possible, by opening containers from time to time, to determine the absorbing activity of the roots at the various levels at any stage in the development of the crop. In these experiments, containers 1.5 to 3 feet in diameter and 2.5 to 5 feet deep were employed. They were placed in trenches, which were then refilled with soil, and crops were planted around the containers in such a manner that the experimental crops were grown under field conditions. Moreover, the containers were filled so that the well-compacted soil at any level occupied the same relative position that it had occupied before removal from the field. To prevent water intake, each container was furnished with an appropriate wooden roof. During a period of two years, 50 containers were used.

In an experiment with barley, for example, it was found that—when the crop was ripe—water had been absorbed from the several 6-inch layers in the following amounts: 20, 19, 16, 14, 12, and 11 percent, respectively (based upon the dry weight of the soil). It also was ascertained, from barley grown in other containers and examined at various intervals, that during the heading and ripening of the grain—a most critical period in plant development—the bulk of absorption was carried on by the younger portions of the roots in the deeper soil.

Corn was found to be an extravagant user of water, absorbing large quantities from the third and fourth foot of soil and smaller amounts from the fifth foot. Potatoes absorbed water to the depths of their root extent, about 2.5 feet, and western wheatgrass and big bluestem (grown from transplanted sods) showed marked absorption to a similar depth.

In other experiments, measured amounts of nitrates (400 parts per 1 million parts of soil) were placed in the soil at various depths. Barley, when 2.5 months old, had removed 286 and 135 parts per million parts of nitrates from the 1- to 1.5-feet and the 1.5- to 2-feet soil levels. It also absorbed 168 ppm. from the 2- to 2.5-feet level, and, at maturity, absorbed 186 ppm. from the 2.5- to 3-feet level. Two native grasses used nitrates from the second and third foot in large amounts, while corn removed 203, 140, and 118 ppm., respectively, from the third, fourth, and fifth foot of soil.

These experiments revealed for the first time the absorption of water and nitrogen at great depths. Thus it was pointed out that the student of environment must consider not only the conditions in the surface soil but also the whole substratum to the depth of root penetration (Weaver, Jean, and Crist, 1922).

A study of the absorption of nutrients from subsoil in relation to crop yield followed in 1924. The same methods as those described were used, and barley was grown in the field in 18 large, watertight, oak barrels. The soil, except in the controls, was fertilized at various levels to 2.5 feet in depth, and in some containers at two levels, with either a nitrate or phosphate fertilizer. A control container without a crop or fertilizer, but sealed at every 6-inch level, was used as a check to determine nitrification or denitrification.

Roots of the control plants often extended to the bottom of the containers, which were 2.5 feet deep, but nitrate fertilizer at any level lessened root depth and greatly increased branching. Both nutrients were absorbed in large quantities at all levels. Although the greatest amount of salts was absorbed in the surface foot, the plants took additional quantities from the deeper levels when it was available. Moreover, absorption of these nutrients below the upper foot materially affected both the quantity and quality of the yield (Crist and Weaver, 1924).

DECAY OF ROOTS AND RHIZOMES

Grass is an excellent preventive of soil depletion and also the best means of soil improvement, as well as man's most efficient weapon against soil erosion. The living, developing root system is the most active part of the grass crop in promoting the formation of soil aggregates. Decaying roots are the source of much of the humus by which the soil particles are cemented into aggregates. These, in turn, are broken up by the mechanical effects of the living roots and are kept from coalescing into clods. The roots of grasses, and therefore the humus produced from them, are distributed widely and rather uniformly throughout the soil. It is believed by some investigators that the pattern of distribution of humus throughout the soil, rather than its quantity, is a main factor in influencing permeability. At least, adding humus to the soil in manures is not nearly as effective in promoting a good structure as the growing of perennial grasses. Upon their death, the roots of grasses distribute the humus in an ideal manner.

Numerous investigations show that young roots decay most rapidly; and they have a higher nitrogen content than older roots. At the early stage of growth, the nitrogen is liberated more rapidly in an available form, namely ammonia, which is readily changed in the soil to nitrate. In contrast to this rapid decay is the persistence of the stele of certain grasses. This part contains much lignin. In the decay of roots in soil, the soluble organic substances are the first to decompose. This is followed by the decomposition of pentosans and celluloses. The lignins, the last of the major plant constituents to decay, turn dark, then black in color, and tend to accumulate in the soil. The quantity of underground plant materials is large throughout the grassland. In the surface 4 inches alone, this ranged from 2.8 to 4 tons per acre in true prairie. Large amounts of food materials are released during a period of several years as underground plant parts gradually undergo decomposition after the death of the plant.

The growing of perennial grasses actually decreases the tendency of the soil to erode even after the soil is again cultivated. For a time, this may be due in part to the effects of undecayed roots and partly to the effects of decaying roots upon promoting soil aggregates or retaining the aggregates already formed. Thus, from a practical viewpoint—as pointed out by Jacks (1944)—a rotation in which land is seeded to grass as soon as the granular structure threatens to disappear, and is kept under grass until a new structure has formed, can to a large extent prevent erosion for a time, even when such widely spaced erosion-conducive crops as corn, tobacco, or cotton are grown.

The rates of decomposition of the underground parts of 12 range grasses were ascertained at Lincoln, Nebraska. Plants of big bluestem, little bluestem, and blue grama, which were growing in prairie on silt loam soil, were killed and left to decay in undisturbed soil. Columns of soil 2 feet in depth and 10 inches in diameter were obtained and were separated into horizontal layers. The layers included the upper 6 inches, the second 6 inches, and the second foot, and one column with each species of grass was examined at the end of each of three years of decay. Plant parts were separated from the soil by repeated washing and screening and then were graded into coarse, medium, and fine materials. The fine materials consisted of fragments of small size which were retained by a screen with a 0.42-millimeter mesh. Control samples from the prairie were similarly treated.

In the columns as a whole, the pattern of decrease (by weight) of coarse materials and increase of fine materials was the same in all cases. In blue grama, the coarse material was 56, 38, 12, and 6 percent of the entire amount that remained (Fig. 50). In this manner the rate of decay of the roots of the three grasses was ascertained. The pattern of decay was the same at all depths for all species, although the percentages were variable. The average percentages for all three depths are shown in Figure 50.

Probably little reduction in the total weight of underground parts occurred the first year. In the second year, little bluestem and blue grama lost 53 and 54 percent, respectively, of their dry weight, compared with the amounts present in the sample examined at the end of the first year. After three years the decrease (from the weight of the preceding year) was 59 and 28 percent. Total decreases in the amounts of residues from the end of the first to the end of the third year were big bluestem 83 percent, little bluestem 81 percent, and blue grama 67 percent. Further studies showed that 49 to 88 percent of the small amount of the remaining materials decomposed during a subsequent period of five months. Thus a small amount of residue, about 80 percent of which was fine, remained after the four periods of decomposition.

Three separate packets of roots and rhizomes, each containing 12 species of grasses, were wrapped in copper-meshed screen and were buried out-of-doors. One lot of material was examined each year. The underground parts of the three species that have already been described decomposed at approximately the same rate as before. In some species the rhizomes decayed more rapidly than the roots.

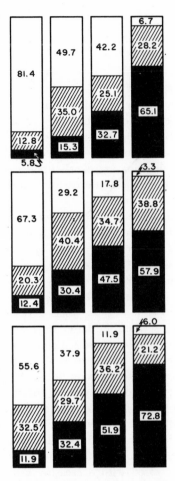

FIG. 50.—Percentages of different grades of big bluestem (upper), little bluestem (middle), and blue grama (lower) in the control (left) and after the first, second, and third year of decay, respectively. Fine material is black; medium material is hatched, and coarse material is white.

Materials of Kentucky bluegrass, Junegrass, needlegrass, and switchgrass decayed most rapidly; only small fragments, if any, were found after three years. Smooth brome was only slightly more resistant to decay. The bluestems, Indian grass, and western wheatgrass decayed somewhat less rapidly. A few roots retained some tensile strength after three years. The grass most resistant to decay was blue grama, but side-oats grama and buffalo grass were closely similar. Much undecayed material remained and some roots of each of these species retained moderate tensile strength after three years. Conversely, Sudan grass and smooth brome, which were grown for one summer, disintegrated as far in one year as many of the perennial grasses had in three years (Weaver, 1947).

Effects of Length of Day and Water Supply

A plant is a very plastic organism. Any factor that influences the development of either the root or the shoot usually profoundly affects the other as well. By 1929, the extensive researches of Garner and Allard (1923), and several others—clearly showed the profound effect the period of daily illumination exerts upon the vegetative and reproductive activities of plants. Little attention, however, had been given to the effect upon the root system.

In our experiments, the relative development of roots and tops of eight species of plants grown under 7-hour and 15-hour daily illumination was ascertained. Red clover, radish, iris, and oats—all long-day plants as regards flowering—developed large tops and proportionately extensive root systems under a 15-hour day. Under short-day illumination, the growth of tops and roots was greatly retarded, and both approximately to the same degree. Their development was similar to that of long-day plants when the latter were only 3.5 weeks old (see Fig. 51). Dahlia, great ragweed, and cosmos—all short-day plants—attained their greatest size and greatest root development under the 15-hour day. Under short-day conditions, the dwarfed tops were furnished with a correspondingly meager absorbing system. Thus, in every species, development of the root system was in direct correlation with the development of tops (Weaver and Himmel, 1929).

It was observed in other experiments that, where dry soil hindered or prevented root elongation, profuse branching occurred to the root tips. Root depth increased with decreasing water content, until the soil became too dry for root growth. Conversely, saturated soil, which may result from excessive irrigation, often causes the death of the roots through lack of aeration (Weaver and Himmel, 1930).

Root Development and Crop Yield under Irrigation

More than half of the surface of the earth receives insufficient precipitation for the most favorable growth of crops, and therefore the best method of making up this deficiency is through the application of water by irrigation. Hence the economical use of irrigation water is one of the chief problems of agriculture in arid regions. Since much more land is available than can be irrigated by the supply of water, even when methods of greatest economy are employed, the welfare of these regions demands that the irrigation water be used as efficiently as

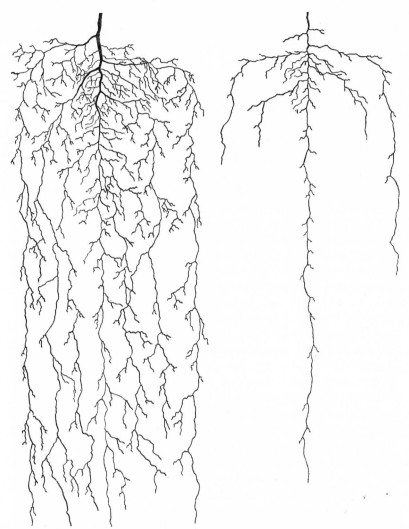

FIG. 51.—Relative development of root system of red clover under 15-hour (left) and 7-hour daily illumination. Depth is about 32 inches.

possible. It should be kept firmly in mind that the criterion of yield under irrigation should not be the maximum production per acre but the acre-inches of water used. Water, not land, is the limiting factor, and we will learn to use the former more effectively as our knowledge of the extent, distribution, and activities of the water-absorbing organ, the root system, increases.

During the seasons of 1922–1923, alfalfa, spring wheat, sugar beets, potatoes, and corn were grown at Greeley, Colorado, in soils that were of very similar nature in texture and chemical composition but varied widely in water content. The growing season was sufficiently long and other conditions also were favorable for plant growth— all except light and unevenly distributed rainfall (12.7 inches mean annual), which makes crop production hazardous. Although about three-fourths of the rainfall occurred during the growing season, much moisture was wasted in light showers and torrential rains, when runoff was very high.

Thirtieth-acre plots of each crop were grown in unirrigated and irrigated sandy loam soil. Consistent and marked differences in the development of the root systems were ascertained as the crops were examined at several periods during their growth. With each crop, it was sought to illustrate the average condition of root development rather than the extreme condition. Extensive study of water relationships was made from the time of planting until the maturity of the crops, and in every instance the root habits were found to be very responsible to variations of this factor. In general, the crops with the most extensive root systems gave the greatest yield.

Water was supplied in these experiments at times designated by the landowner, a man of many years' experience in irrigation and whose judgment in such matters has been attested by highly successful crop production. The water content of the soil was ascertained immediately preceding and following irrigation of the readily permeable soil by taking numerous soil samples at 1-foot intervals to depths of 3 feet.

Roots of Turkestan alfalfa in the dry hardland differed from those in the irrigated soil by pursuing a much more tortuous course— probably because of the difficulty of penetration—and by having longer major branches, which were especially abundant in the first foot and which turned downward more abruptly. In the hardland, the earlier development of the plant's major branch roots often reached as great a depth as those of the taproot. Likewise, secondary and tertiary branches were much longer, owing to the slower growth of the taproot and largest laterals, and branches on them occurred much nearer the tip (Fig. 52). The dry soil had a very retarding effect upon the growth of tubercles, and none occurred at the end of the growing season, although they were abundant to 4 feet in depth in irrigated soil. Under irrigation, alfalfa was more than 3 times as tall as alfalfa on dry land, and gave an excellent yield.

Roots of Marquis spring wheat in dry land, because of a lack of moisture in the second foot, spread more widely in the surface soil, where the moisture was replenished by the heavier summer showers. Many of the roots died of drought and growth was greatly retarded.

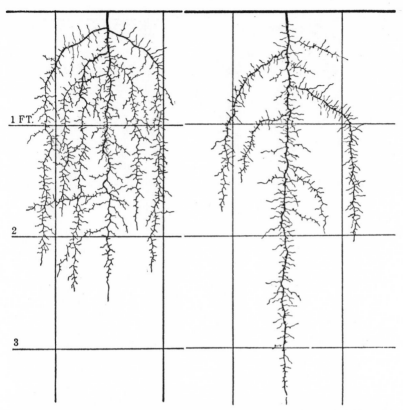

FIG. 52.—Roots of alfalfa plants 3 months old, grown in dry land (left) and in irrigated soil. Note the greater number of primary and secondary branches in the dry land species.

The wheat plants that finally turned downward spread much more widely than those in the plots with a moist subsoil (see Fig. 53). Although the number of primary branches was no greater in the dry land, they were longer, and occurred nearer the root tips; and the tertiary branches were much more numerous. Dry-land wheat grew but little after the middle of June, having extended its roots into the still rather dry second foot of soil to a level of only 2 feet, as contrasted to 4.3 feet in the irrigated plot. The height of the crop in the dry soil was 15 inches and the yield was 3 bushels per acre. In the irrigated soil the height was 43 inches and the yield was 29 bushels per acre.

In dry land, Klein-Wanzleben sugar beets had their normally deeply penetrating taproots limited in their early development, because of dry subsoil, to one-half the depth attained in moist soil.

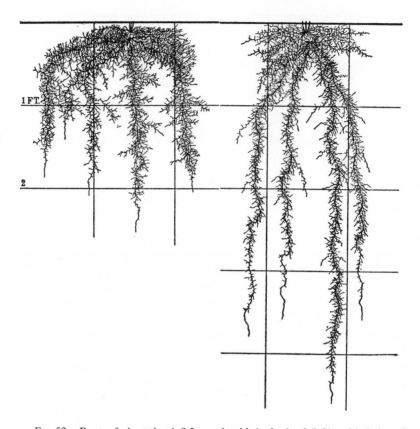

Fig. 53.—Roots of wheat that is 2.5 months old: in dry land (left) and in irrigated soil.

Throughout the season they were 1 to 1.5 feet shorter (Fig. 54). By mid-September the dry-land beets, which had suffered severely from drought, had only about one-fourth the leaf area of the irrigated plants, and the area occupied by the roots was much less. The surface roots, stimulated by water from showers, had reached a lateral spread that exceeded the spread in moist soil, and abundant deeper-seated major branches had penetrated the dry subsoil 2 to 3 feet. As in alfalfa, they did not spread widely (seldom more than several inches, as compared with 1.5 to 2 feet in moist soil) but turned downward abruptly. A maximum depth of 4.5 feet was attained, but roots in moist soil extended 6 feet deep. Yields in dry and irrigated soils were 2.5 and 22.5 tons per acre, respectively.

Late in June, Bliss's triumph potatoes had smaller tops but much more extensive root systems in the dry land than in irrigated soil.

Moreover, the root branches were much more abundant and longer. Because of drought, little further growth occurred, either above or below ground, in the dry-land potatoes, whose general root depth was about 2 feet. The irrigated crop was rooted mostly in the surface 16 inches of soil, where the lateral roots spread nearly 2 feet on all sides of the plant. Tubers were abundant in both plots but were larger in the irrigated plot. Yield at maturity in dry land was at the rate of 19 bushels per acre, all tubers being small. In the irrigated plot it was 160 bushels per acre.

Yellow dent corn, Minnesota No. 13, a variety that is considered best adapted for this region, was planted on May 10. There were marked differences in the root habit, even when plants were only 6 weeks old. Roots in dry land deviated from the normal, shallow, widely spreading type, and more of these roots grew obliquely or directly downward. The number and length of branches increased greatly in the dry soil. By midsummer the dry-land corn had almost completed its growth, both above and below ground. Drought and excessive transpiration had stimulated a marked root growth. The widest lateral spread of

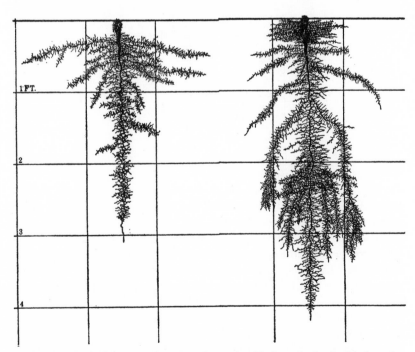

FIG. 54.—Sugar beets that are about 3 months old: from dry land with dry subsoil (left) and from fully irrigated soil.

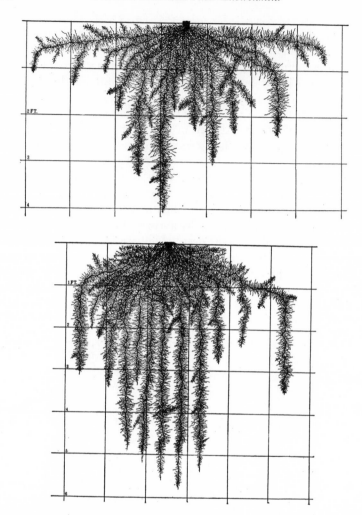

Fig. 55.—Roots of dry-land corn on Sept. 12 (above) and roots of corn in fully irrigated soil on Sept. 13.

shallow roots (about 3.5 feet) and the most profound branching now occurred here, but the general depth (about 3 feet) was limited by dry soil. In the irrigated plots, with plants 1 foot taller, almost the entire root system (as is usual for corn) was limited to the surface 16 inches of soil.

The mature root system in dry land, like the tops, showed little further growth after midsummer, and the yield was only 25 bushels per acre. Plants in the irrigated soil had extended their roots well into the fifth and sixth foot, and many of the horizontal laterals also had turned downward and had penetrated deeply. Here the yield was 102 bushels per acre (Fig. 55).

It is clear from the preceding account that the roots of the crops that were employed responded readily to the changed environment, which consisted chiefly of differences in the available water supply and air content. Also, there is a striking correlation between the growth of underground and aerial plant parts. Other extensive experiments have shown that the root habits of a few species seem to be controlled largely by heredity and that they show little plasticity under changed conditions; the behavior in most plants depends upon the operation of such factors as water content, aeration, and nutrients. Marked changes of root habit may be brought about by various methods of tillage, the use of fertilizers, crop rotations, cover crops, intercropping, and especially by irrigation and drainage. Examples are given in this book (Jean and Weaver, 1924).

The development of each species in lightly irrigated plots also was studied and the effects on root development were ascertained. Moreover, the second year (1923) of these experiments witnessed a very wet summer, which considerably modified root habits in dry land. These interesting differences in unirrigated soil cannot be recorded here.

VI.

Studies on Plant Competition

The struggle for existence among plants is almost always between each plant and its habitat. The habitat is changed in consequence of the demands made upon it by other plants. Competition is essentially a decrease in the amount of water, nutrients, or light available for each individual plant. It always occurs where two or more plants make demands for light, nutrients, or water in excess of the supply. Competition begins when plants become closely grouped, as in prairie, chaparral, and forest, and it continues even when vegetation is stabilized.

In a comprehensive monograph on plant competition (Clements, Weaver, and Hanson, 1929), the scope of the concept of plant competition was given adequate description. Like other primary concepts in ecology, the idea of competition has gradually emerged from the general experience of mankind. It must have appeared long before it was recorded, just as the record must have preceded a name for the process itself. As the most striking consequence of the grouping of organisms into communities, it was perhaps first clearly perceived in the case of the forest, but its manifestations were doubtless evident long before in human societies. Certainly, the demand for an essential factor in excess of the amount afforded by the environment must have early engaged the attention of thoughtful men. The existence of the same process in plants and animals could hardly have escaped notice, but the formulation of the idea had to await the beginning of scientific forestry and agriculture.

The earlier views of the concept of competition—between 1798 and 1904—are discussed and the list of experimental studies between 1905 and 1928 is given. Studies of competition and succession by various researchers are described, and—especially—competition in forest and cultivated fields in given adequate attention. A wealth of information is included in this 30-page review. Most of the following data are taken from this monograph.

COMPETITION BETWEEN GRASSES AND BETWEEN GRASSES AND FORBS

Extensive studies were made in lowland prairie at Lincoln, Nebraska, from 1924 to 1926. In competition cultures in the prairie, dominant tall grasses, such as switchgrass (*Panicum virgatum*), nodding wild-rye (*Elymus canadensis*), and big bluestem (*Andropogon gerardi*), were grown in competition with others from seedling to maturity (see Fig. 56). A monthly record of results was obtained. Tall grasses were pitted against short grasses, such as blue grama (*Bouteloua gracilis*) and hairy grama (*B. hirsuta*). Dominant grasses were made to compete with various prairie forbs, and in other quadrats with ruderals. Similar studies were made in adjacent upland prairie.

Competition for water and nutrients involved a knowledge of root systems. A very intimate knowledge of the growth of both tops and roots was acquired, which helped greatly in understanding prairie vegetation. Not only a knowledge of which species was the better competitor was obtained but also the reasons for its success. Many pages were necessary to record these studies, but only a few examples of the results can be given.

Competition between little bluestem and a long-lived perennial forb, false boneset (*Kuhnia eupatorioides*), in the fall of the first year (1924) is shown in Figure 57. The more rapid growth and greater spread of false boneset gave it a distinct advantage the first year, in spite of the fact that little bluestem outnumbered it 4 to 1. It dominated the quadrat, regardless of the fact that 32 percent succumbed under the intense competition and none made a normal growth. However, they attained a height of nearly 1 foot, and flowered and seeded abundantly. Kuhnia was able to obtain the larger share of soil

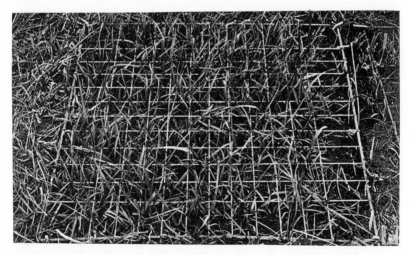

Fig. 56.—Competition culture of *Elymus canadensis* and *Panicum virgatum* in low prairie. The small divisions are square inches.

Fig. 57.—Competition culture of *Khunia eupatorioides* and *Andropogon scoparius* at the end of the first summer.

water by virtue of its deeper root system, and this permitted it to over-top the grass at all times, three-fourths of the grass plants being badly suppressed.

During the second season the roots of the forb extended to a depth of 4 to 5 feet, and quite beyond those of the grass. This enabled it to absorb vigorously below the general competitive level and to place its leafy tops at a constant advantage with respect to light. As a consequence, the initial lead due to better absorption was converted into an even more decisive gain as to light, and shading proved to be the most critical factor for the grass.

The third season was one of severe drought. The weaker plants of little bluestem had been eliminated by shading and winter-killing, and the survivors were better fitted for competition. They formed a dense sod and developed some flower stalks. Less than a third of the plants of Kuhnia were sufficiently thrifty to bloom, and all had assumed a position of subdominance with respect to the bluestem.

In the fourth season, both species were in good condition and the quadrat exhibited the usual structure of true prairie, in which the subdominant forbs maintain themselves in greater or less abundance in accordance with the rainfall and the consequent growth of the dominant grasses.

Big bluestem, in separate plots, was matched against two ruderals, *Amaranthus retroflexus* (redroot) and *Ambrosia trifida* (giant ragweed). The grass germinated about 4 times as well as either competitor, and possessed the further advantage of potential height and better root system. For a time the competition between the bluestem and the redroot was keen both for water and light, but lack of water and nutrients was chiefly responsible for the marked suppression of this ruderal. The rapid growth of ragweed enabled it to take the lead and hold it during the first summer, chiefly in consequence of the reaction of the tall, spreading stem and broad leaves upon light. Unlike the redroot, some of the seed that were produced germinated the second spring, but the seedlings were unable to compete with the better-established grass and the final fate of the two weeds was the same.

Shrub Communities; Ecotone between Forest and Prairie

The place of transition from prairie to woodland is clearly marked by the trees along the forest border. This is considered a tension zone between grassland and forest, and in general it results from a change in the climatic processes. But the success of tree plantations in the prairie has led to the assumption that even the true prairie owes its persistence to fire and that the climatic relations are not controlling. The fact has been overlooked that such groves have been artificially aided in a number of ways: by the destruction of the grass cover, by the use of mulches or actual watering, and by the employment of young trees tall enough to escape overshading. In short, the most critical time in the whole process, that of germination and establishment, is avoided by the use of transplants, and the physical factors and competitive relationships are profoundly modified to the advantage of the tree.

The forests along the Missouri River in eastern Nebraska have been thoroughly studied and will be described in the following chapter, as well as the woodlands along streams. The close relationship of the communities of trees to water content of soil, wind, and humidity have been correlated. This was accomplished by plotting the number and kinds of trees in 10-meter-wide and very long transects that extended over the hills and bluffs of southeastern Nebraska. The most mesic trees, such as linden (*Tilia americana*) and red oak (*Quercus borealis* var. *maxima*), occurred mostly in the ravines and on lower, moist slopes. Black oak (*Q. velutina*) and shellbark hickory (*Carya ovata*) occurred on intermediate sites, and bur oak (*Quercus macrocarpa*) and bitternut hickory (*Carya cordiformis*) occurred on the drier slopes and hilltops.

About 40 species of shrubs and vines occurred as undergrowth; some of them grew in deep shade, but most of them grew in the more open, better-lighted, but drier forest types.

Many of these shrubs extend for some distance beyond the forest border and form a more-or-less-complete area of transition from forest to prairie. This may be only a few yards in width in places, but it usually is several to many rods wide, and some species of shrubs extend along streams and ravines far into the grassland. In fact, extension of trees into true prairie is almost always preceded by chaparral. A list of the more common shrubs and vines follows.

SHRUBS

Amelanchier canadensis	*Rhamnus lanceolata*
Amorpha fragrans	*Rhus glabra*
Cornus drummondi	*Ribes missouriense*
Cornus stolonifera	*Rubus occidentalis*
Corylus americana	*Sambucus canadensis*
Euonymus atropurpureus	*Symphoricarpos occidentalis*
Prunus americana	*Symphoricarpos orbiculatus*
Prunus virginiana	*Zanthoxylum americanum*

VINES

Celastrus scandens	*Rhus toxicodendron*
Clematis virginiana	*Smilax hispida*
Lonicera.diocia	*Vitis vulpina*
Parthenocissus quinquefolia	

The most important invaders of grassland are buckbrush and wolfberry (*Symphoricarpos orbiculatus* and *S. occidentalis*, respectively), smooth sumac (*Rhus glabra*), and hazel (*Corylus americana*) (Figs. 58, 59). The first two species occur in lowlands and ravines throughout true prairie, and the last species is restricted to the edges of the deciduous forest. All along the forest's border, from Minnesota southward to Oklahoma, hazel comes in direct contact and competition with the prairie grasses.

These shrubs are tall plants, which overtop the bluestems, and, being deeply rooted, they often extend downward as deep as—or deeper than—the roots of prairie grasses. Shrubs spread widely by various vegetative methods—stolons, rhizomes, and root offshoots—and then shade out the grasses. Sumac usually pioneers in grassland, often on the slopes of hills. Buckbrush and wolfberry usually become dominant in less-exposed situations, and hazel endures more shading by trees and is the least xerophytic of the lot.

FIG. 58.—Coralberry (*Symphoricarpos*) invading a pasture in eastern Nebraska.

FIG. 59.—Sumac (*Rhus glabra*) in a ravine in a large, mowed prairie in eastern Nebraska.

COMPETITION BETWEEN SHRUBS AND GRASS

The fascinating problem of the relationship of the forest edge to the grassland has been much debated on the basis of general observation but has never before been attacked in a comprehensive manner. In 1915, study was begun by Weaver and Thiel (1917) near Minneapolis, Minnesota, where a very large prairie was separated from an oak forest by a transitional area of shrubs, chiefly species of Corylus, Rhus, and Symphoricarpos.

It was ascertained that, in general, the soil's water content during the growing season was greatest in the forest, intermediate in shrub thicket, and least in the prairie. Moreover, daily evaporation was lowest in the forest and highest in the prairie, the differences being great and continuous throughout the growing season. Phytometers, consisting of groups of bur oak seedlings placed in prairie, shrub, and forest, revealed a much higher rate of transpiration in the prairie— where temperatures were higher, light was more intense, and wind movement was greater—than in either the shrub or forest habitats. Transpiration in prairie was 3 times greater than in the hazel thicket, and about 10 times greater than in the oak forest. There was a general correlation between the evaporating power of the air and the amount of transpiration. In brief, the amount of evaporation in the prairie, coupled with the relatively low water content of the soil, is sufficient cause for the xerophytic character of the vegetation.

Similar studies were pursued in prairies, sumac thickets, and flood-plain forests in southeastern Nebraska for several years (Weaver and Thiel, 1917; Pool, Weaver, and Jean, 1918). Environmental conditions were continuously measured by means of batteries of field instruments at the several stations, as well as by phytometers. Later, similar studies were made by Aikman (1927) in various habitats in his studies of forests. Measurements in numerous chaparral and prairie areas over a period of years revealed that the areas with shrubs are characterized by an increased water content of soil, higher humidity, lower soil and air temperatures in summer, decreased wind velocity, reduced evaporation, and decreased light intensity.

Ravines in true prairie sometimes contain thickets of one or more species of shrubs that constitute the marginal chaparral of the decid-uous forest. Where they are extensive and continuous, they are probably relicts of a former forest, but they may also arise from recent invasions due to birds—especially at times when the hold of the grass sod is weakened by flooding.

A thicket of *Rhus glabra*, with an area of about 500 square feet occurred in a large prairie a few miles north of Lincoln. It extended upward from a ravine on the lower part of a southwest slope. On its lower edge it was in contact with a community of prairie cordgrass and on its upper edge it was spreading into the prairie in spite of annual mowing. This thicket had been known for 30 years, and it was believed its lack of farther spreading resulted from annual mowing. Consequently, a considerable area bordering the thicket was enclosed to prevent this disturbance.

The number of shoots in the marginal area of uninvaded prairie was 174, of which 70 were new in 1924. In the fall of 1926 the number of shoots had decreased by 28 percent. In 1927, further examination showed conclusively that the sumac community was not only making no progress up the slope into the prairie but was steadily losing ground in competition with the unmowed grasses. The height growth in prairie was very small compared with the taller growth on the lower slope. This resulted in a flat-top group of shrubs. Although an outlier was occasionally found several yards beyond the exclosure, it exerted no effect of any consequence upon the grasses. Thus one of the most aggressive members of the shrub community showed its inability to advance into true prairie. This is decisive evidence of the nature of the climatic control of the corresponding climax.

Later examination along old roads through the prairie and across ravines with sumac—areas that had been abandoned for many years— also clearly showed the inability of sumac or other shrubs to extend far up the slopes.

A study in the ecotone between prairie and forest extended over a period of several years at Weeping Water, about 30 miles east of Lincoln. Here the rainfall was about 2 inches higher than the 28 inches at Lincoln. The Weeping Water River had cut a canyon more than 100 feet deep in the Pennsylvanian limestone, and this—with its similarly cool, moist lateral (Cascade Creek)—provided a refuge for the deciduous forest as it underwent shrinkage during the last major dry climatic phase. Trees and shrubs constituted a cover of woodland about a mile wide, with alternes of prairie only on the dry, windy hilltops and southwest slopes. The upper and drier margins of the woodland were fringed with chaparral, which also extended beneath the trees as a more-or-less-interrupted layer.

In the shrub community, Corylus exhibited a distinct preference for the more sheltered and moist areas, though it sometimes occurred on protected hilltops at the edge of the grassland. Associated species were *Cornus drummondi, Ribes gracile, Zanthoxylum americanum,* and *Rhamnus lanceolata,* and such lianes as *Celastrus scandens* and *Smilax hispida.* By means of their excellent methods, the shrubs appeared to be spreading and overtopping and shading out the bluestems and other prairie grasses.

Two areas in which the shrubs and grasses were competing were protected by fencing to prevent grazing. The areas were charted in May, 1924 and 1925, and in August, 1926. The shrubs were spreading in all directions and clearly encroaching upon the grasses. The shrubs exerted a decisive reaction upon the grasses, not only by shading but also by the accumulation of their fallen leaves and other debris.

The prairie grasses gradually yielded to *Poa pratensis*, where they were shaded for a half-day or more, and when the light intensity was reduced to 10 percent. When the light intensity approached 3 percent as a consequence of the increasing density of the hazel, the bluegrass also disappeared; the accumulation of litter played a part in this result. Several clumps of *Andropogon scoparius* were plotted and their fate was followed in close detail, as shown in Figure 60.

In 1924 a selected bunch of *A. scoparius* was sufficiently lighted to make a good growth and bear flower stalks. By 1925 the hazel had reached it, and a large branch extended well over the grass. However, it still bore inflorescenses, except on the two shaded sides (Fig. 60). By 1926 this bunch had become further overshaded and invaded; no flower stalks developed and there were only a few green leaves. A shoot of hazel had arisen from a rhizome that extended beyond the grass, and a young stalk of Celastrus had appeared on the other side of the bunch. On July 10, 1927, only a few weak leaves were found, and the steady advance of hazel left no doubt of the immediate outcome. Two clumps of *Andropogon gerardi* were likewise traced in detail; first they ceased flowering, and then they died.

This encroachment by the shrubs was only temporary. After a series of dry years, they suffered almost total loss. All woody plants in the shrub area were swept away and the prairie came into full possession of the hilltop and the upper, dry southwest slopes it had previously occupied.

This study and other similar studies should be carried on through even longer periods of time, but the mass of evidence accumulated from studies in the prairie-forest border lead to the following conclusion. The slow advance of chaparral and forest will be hastened by the wet climatic phase and retarded by the dry phase, or even converted into a retreat. (See Chapter 7, "Death of Trees during the Great Drought"). It appears fairly certain that there can be no final victory for either; there can be only periods of varying duration in which prairie or forest holds the ground won by the favor of the changing climatic cycle.

Fig. 60.—Competition between *Corylus americana* and *Andropogon scoparius*: Aug. 21, 1925 (*A*), July 15, 1926 (*B*), extinction of the grass, July 10, 1927 (*C*).

COMPETITION BETWEEN TREES AND GRASS

An experimental planting of trees in low prairie at Lincoln was begun in 1924 and extended over a period of four years. Five species of trees that are common on flood plains of streams and that are often grown for shade and windbreaks were used: *Acer saccharinum, Ulmus americana, Gleditsia triacanthos, A. negundo,* and *Fraxinus lanceolata.* Seeds were employed for the first three, and seedling transplants with the last two. The plan was to secure four degrees of competition with the grasses, in increasing severity. In every instance it was first necessary to break the hold of big bluestem, switchgrass, and other lowland grasses. Each square foot of sod was bound with 50 to 60 feet of intertwined rhizomes and vast networks of roots to great depths. Hence four long parallel trenches were made in the sod, 4 inches wide and 4 inches deep. These were filled with prairie soil that was free from roots and other debris, and the soil was firmed in place. Then seed was sown or the seedlings were transplanted. Water was added from time to time to ensure germination and establishment.

In the first row the sod was overturned to a depth of 4 inches, to a distance of 6 inches on either side, and it was thoroughly pulverized to constitute a good mulch. By frequent shallow hoeing this denuded area was kept free from invaders during the following seasons, while the overhanging grasses were repeatedly clipped to ensure good lighting and thus confine the competition to the soil. Even here, for a time at least, there was little or no competition for water or nutrients since the roots of the dominants of low prairie spread but little and penetrate deeply and almost vertically.

The grass along the second trench was kept clipped to the ground for a distance of 6 inches on both sides. Thus there was almost no competition for light, and the demands of the clipped row and the competition for water and nutrients were not severe. Along the third trench, grasses and trees were watered freely from time to time during the first year, and especially in periods of drought. Since sufficient water was present at all times, competition in this row was for light, and to some extent for nutrients. Trees in the fourth trench were flanked by the prairie grasses and were entirely unaided in competition with them (see Fig. 61).

During the spring and summer of 1924, conditions for growth were very favorable, although rainfall during August and September was slight. The trees in the mulched row made much better growth than in the clipped row. Those in the other rows did much more poorly; there was little difference between the watered and unaided rows as a whole, and the mortality among the trees increased with the

Fig. 61.—Growth of boxelder at end of first summer in mulched, clipped, and un-aided rows.

degree of competition. The general growth and height was greatest in the mulched row, next-greatest in the clipped row, and least in the two remaining rows. Watering caused the grasses to grow more vigorously, with the consequence that the trees received less light in the unaided row. In both rows they became surrounded by a stand of grass 3.5 to 4 feet tall.

During the next year (1925)—with *Acer saccharinum* as an example—the average height in inches in the four rows was, in order, 21.7, 9.1, 6.5, and 7 inches, and the average diameter of stems was 9, 4, 2.5, and 2 millimeters. The number of trees that died during the summer were 1, 12, 13, and 12, in the same sequence.

The dry weather resulted in much winter-killing, the spring of 1926 was late and cold, and the summer was dry until mid-August. Mortality was least in the mulched row, next-lowest in the clipped and unaided rows, and greatest in the watered row. Many of the twigs were frozen and the terminal buds or a few inches of the stem were dead.

Even during the driest part of the summer, trees in the mulched row generally were in good condition, although there was some folding and rolling of leaves. In the clipped row, some trees were dying. By early August the condition of the trees had become somewhat worse; some of those in the unaided and watered rows were badly wilted. By fall, the average loss of all trees in the mulched row was 31 percent. In the other rows, in sequence, the losses were 62, 79, and 92 percent. In the mulched row, losses by species were as follows: Gleditsia, 13 percent; Ulmus, 14 percent; *Acer negundo* 20 percent; *A. saccharinum* 40 percent; and Fraxinus, 67 percent. As measured by height and dry weight, *Acer negundo* made the best growth.

The water content of the soil was ascertained under each condition for growth almost continuously during the three years, but it cannot be recorded here. The best explanation for the success of some species and for the failure of others is to be found in the development of the root systems. In fact, the rate and degree of development and the fineness of branching were regularly correlated with survival.

Root Development under Competition

The root system of each species under the various degrees of competition was examined in careful detail during August of the third season of growth, 1926. A long trench, 3 feet wide and 6 or more feet deep, was dug parallel to the mulched row of trees but at some distance from it. The root systems were then excavated by digging into the wall of the trench with special care. After several individuals of each species had been examined and a typical root system had been drawn, the walls of the trench were cut back until the second, or clipped, row was encountered. Smaller trenches were employed in the excavation of plants from the watered and unaided rows; the trees in both of these rows were so similar in size that one lot from either sufficed for both. In the drawings, the roots are shown in a single plane, while a single representative is drawn to the natural size, and all of the other drawings were made to scale.

The root system of the honey locust was the most profusely branched of all, which explains the ability of this tree to endure drought (Fig. 62). As with all the trees in the mulched row, the roots exhibited a marked preference for the mulched area; the maximum lateral development always was found here and not in the sod on either side. Roots frequently extended into the sod and then curved back again into the loose, moist soil beneath the mulch, though many also came into competition with the grass roots. Seldom did a root that originated in the first foot turn downward, but several of those from the second foot took this course, and long branches often ran parallel with the taproot. These, in turn, were so profusely branched that the absorptive area in the deeper soil was considerably increased.

The root system in the clipped row closely resembled that of the mulched trees, but it was much less extensive. The diameter of the tap-

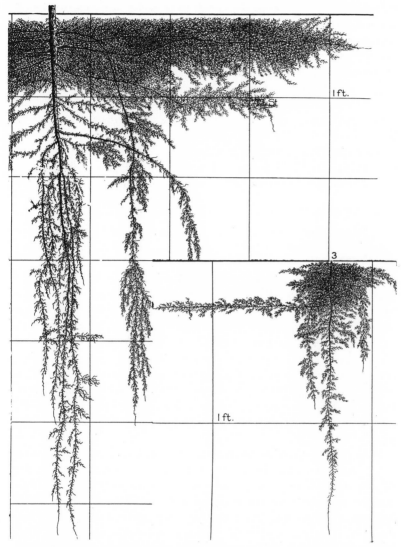

Fig. 62.—Roots of honey locust (*Gleditsia triacanthos*) from mulched and unaided rows.

root was only one-fourth as great as that in the mulched row. Less than half as many horizontal branches were produced, and their length was not 3.5 to 4.5 feet up and down the row but about 2.5 feet. The surface foot was not nearly as well occupied by mats of branches. The thin taproot, or one of its branches, extended downward 5.5 feet, but the whole root system below 1 foot had fewer and shorter major branches, and these were clothed with fewer and much shorter rootlets.

In the clipped row, the taproot was less than 2 feet deep and—except for a single, foot-long, horizontal lateral—other branches at any depth scarcely exceeded one-half foot in length (see Fig. 62).

These and many extensive, similar experiments with trees and shrubs in upland and lowland prairie point out the extreme difficulties of their becoming established in grassland because of the competition for water in a prairie habitat.

COMPETITION AMONG CROP PLANTS

The struggle for existence in the plant world is waged between each plant and its habitat, the latter being changed by competition in consequence of the demands made upon it by other plants. Competition

FIG. 63.—Representative plants, showing development of Sudan grass in (left to right) the 3N, 2N, N, 1/2N, and 1/4N plots on July 17.

Fig. 64.—Sunflowers, of average size, from (left to right) the $2N$, N, $1/2N$, and $1/4N$ plots. The largest sunflower is 64 inches tall.

always occurs if two or more plants make demands for the habitat factors—light, nutrients, or water—in excess of the supply. It means, essentially, a decrease in the amount of water, nutrients, or light that is available for each individual.

Extensive studies of competition in cultivated fields at Lincoln were conducted over a period of several years, especially on sunflowers and spring wheat. Some of these results became widely known after they were discussed and pictured in an 18-page story of competition in *Plant Ecology* (Weaver and Clements, 1938), which is now available in several foreign languages, as well as in English.

The degree of plant competition was measured and recorded in a unique manner by Peralta (1935). Plots of Sudan grass (*Holcus sorghum sudanensis*) were planted at the normal rate (N), 22 pounds per acre, at one-quarter and one-half this rate, and also at twice and three times this rate ($2N$ and $3N$). Seed was planted on fertile lowland soil on May 22, 1933. On June 5, when the Sudan grass was well established, 30 sunflower seeds were planted in each plot and widely spaced. Germination was prompt, and growth at first was vigorous, but there was no time during the summer that the grass was overtopped by the sunflowers, even in the thinnest plantings.

Representative plants that showed the development of Sudan grass in $3N$, $2N$, N, $1/2N$, and $1/4N$ plots on July 17 are shown in Figure 63. The relative heights of plants from the several plots is indicated by the meter rule. The dwarfing of stature and the restricted production of tillers is attributed directly to the decrease in light and water, and probably—in part—to the decrease in nutrients. In Figure 64, the increasingly greater development of all parts of the sunflowers with an increase in water and light—and as the Sudan grass became thinner—clearly shows the harmful effects of competition upon the individual. All sunflowers in the $3N$ plots had succumbed, and most of those in the $2N$ plots were dead. Height increased gradually, from 41 to 64 inches, and the total leaf area per plant increased from 2.4 to 112 square decimeters, respectively.

VII.

Studies in Woodlands

Only about 3 percent of Nebraska is forested. The main areas are deciduous forest, mostly in the southeast, and coniferous forest, in the northwest. Woodlands, however, are scattered widely along the flood plains of many rivers and streams. The state, as a whole, slopes eastward between 8.5 to 9 feet per mile, from an elevation of about 4,000 feet or more to the Missouri River. In the northeast, the altitude is nearly 2,000 feet, but it is only 825 feet in the extreme southeast. Consequently, the great rivers—the Niobrara, Elkhorn, and Platte—flow eastward or southeastward into the Missouri, which bounds the state on the east. It is along these rivers and their numerous tributaries that the flood-plain forests occur.

Trees along Streams

The occurrence of trees along streams was studied from the streams' sources to their union with rivers (Weaver, Hanson, and Aikman, 1925). This was accomplished by means of the transect method, which consisted of the examination of a mile or more of vegetation at intervals along the streams' course. An example follows.

An initial transect was made along the Weeping Water, in eastern Nebraska, at its source. Here the grass-covered land sloped to form a broad valley. In its lower part, an intermittent stream—dry except after heavy showers—began to cut into the sod to form a channel. As the channel deepened and widened along the first few miles of its course, it presented a habitat where wind-blown seeds of willow and cottonwood germinated and where the seedlings developed. This occurred as soon as the soil was bared; then all along the stream, as isolated individuals or intermittent clumps, *Salix amygdaloides, S. nigra,* or *Populus deltoides* appeared.

A few miles farther down the valley was a spring-fed stream that had ceased to be intermittent. Joined by other streams, it had cut its banks 10 feet or more wide, and was 6 to 8 feet deep. Wind-protected, sloping banks are favorable sites for certain shrubs. Chief among these are coralberry (*Symphoricarpos orbiculatus*), wolfberry (*S. occidentalis*), elder (*Sambucus canadensis*), and indigo bush (*Amorpha fruticosa*). Also present, in smaller numbers, are smooth sumac (*Rhus glabra*), wild gooseberry (*Ribes missouriense*), frost grape (*Vitis vulpina*), and bittersweet (*Celastrus scandens*). Willows, 8 to 15 inches in diameter, and cottonwoods, a few of them 3 feet thick, were much older and larger than those upstream. Green ash (*Fraxinus lanceolata*) and chokecherry (*Prunus virginiana*) were represented by one or two small trees.

Farther downstream an increasingly large number of trees and shrubs occurred, as well as the beginning of their separation into different habitats. Fine, large trees of red or slippery elm (*Ulmus rubra*) and white or American elm (*U. americana*) were most abundant on the banks. Flood-plain species were boxelder (*Acer negunda*), green ash, hackberry (*Celtis occidentalis*), honey locust (*Gleditsia triacanthos*), and black walnut (*Juglans nigra*) (see Fig. 65). Some of these, of course, grew near or among the elms as well as on the base of the slope. A scattered growth of young bur oak, (*Quercus macrocarpa*), usually with diameters of 5 inches or less, grew on protected north-facing slopes and on the sides of lateral ravines.

Aside from the shrubs and vines found upstream, which occurred much more abundantly here, several other species had migrated upward from the Missouri River. These were rough-leaf dogwood (*Cornus drummondi*), black raspberry (*Rubus occidentalis*), burning bush (*Euonymus atropurpureus*), greenbrier (*Smilax hispida*), poison ivy (*Rhus radicans*), Virginia creeper (*Parthenocissus quinquefolia*), buckthorn (*Rhamnus lanceolata*), and virgin's bower (*Clematis virginiana*). Most of these were scattered widely over the flood plain. Other transects were made to the Missouri River.

Continued study of the beginning of woody vegetation along the streams over this central prairie region—along the branches of the Nodaway, Tarkio, Boyer, Elkhorn, and other rivers in Iowa and Nebraska—confirmed the general sequence described. Pioneer trees at the stream sources are those with light, wind-blown seeds, such as willows. They usually appear soon after the sod is weakened by erosion. After a suitable habitat is available, other species with wind-blown seeds occur, such as boxelder, elm, and ash. The pioneer shrubs and vines— elder, coralberry, bittersweet, grape and others—have showy, edible fruits that are carried by birds. This early stage in woodland development is represented for considerable distances along nearly all of the small tributaries. It is especially pronounced in the eastern part of the area, as well as northward—along the tributaries of such rivers as the Elkhorn and the Big and Little Blue, which extend far into the plains.

When a stream develops a flood plain with wide, protecting banks,

FIG. 65.—View in flood-plain forest near Weeping Water, Neb. The trees are mostly boxelder, green ash, walnut, white elm, and hackberry. Photo was taken in 1929.

large fruits such as walnut, bur oak, hickory, hazel, etc., are carried upstream by various animals, especially by timber squirrels. Thus as soon as a suitable habitat is provided in wind-protected places along the stream, trees and shrubs—in the absence of recurring fires—may replace the prairie grasses. On unsheltered and windswept banks of prairie streams, however, and on those far from large rivers, little or no woody vegetation may occur.

Far down most of the streams in the southeastern part of the area, flood-plain forest is separated from bur oak and its border of shrubs by more mesic types of forest, especially by red oak and linden (Weaver, Hanson, and Aikman, 1925).

VEGETATION OF LARGE BOTTOMLANDS

Over the flood plains or bottomlands of the Missouri, Platte, and lesser rivers, and in addition to trees and shrubs which more or less border the rivers, the plants of swamps and marshes and of three communities of grassland occur (Weaver, 1960). Swamp plants live in the abundant, shallow lakes and ponds or on the margins of deeper ones. Sedges and rushes occupy marshes whose soil is very wet in spring and early summer, when it may be covered with several inches of water. Coarse grasses, such as prairie cordgrass and switchgrass, occupy large areas of soil on higher ground than the marshes, but the largest grassland community is that of big bluestem, which prevails mostly on second bottomlands whose soil is well drained.

The flood plain of the Missouri is a vast area of level lowland, and only when one observes it closely are slight changes in elevation noticeable. From the river's margin, a distant view is obscured either by steep, wooded bluffs on one side or—on the other—by trees, shrubs, and vines of the flood-plain forest. Since the apparently nearly level topography varies only a few feet below or above the general level, many habitats and various communities seem to the casual observer to be quite intermixed. Actually, as will be shown, there is a rather definite relationship between the kind of vegetation and the type of habitat it occupies.

The flood plain of the Missouri River is 0.5 to 1.5 miles wide between South Dakota and Nebraska. In places southward, it is 17 miles wide, but varies from 8 to 14 miles near the Kansas line. The larger part of the flood plain in this area—perhaps 80 percent—is in Iowa. It is bordered on the east mostly by a broken line of bluffs, but in places only by hills. The channel often lies close to the steep bluffs that form the western boundary of the valley. Bluffs border many of the larger streams in part, but often the valleys are bounded only by hills. Each river has its flood plain, and many smaller streams meander through a plain that is 0.5 to 3 miles wide before they enter the larger plain of the Missouri.

The Missouri River bottomland is a nearly level plain of alluvial sediments. The topography is uniformly level, except in relatively small areas where low ridges and shallow swales and slopes along the edges of old channels cause gentle undulations. Infrequently, deposits of sand have been blown into low hills in places. Usually the land has slopes of less than 2 to 3 percent. Meander scars are characteristic features of flood plains.

Flood-plain forests along the larger rivers consist very largely of cottonwoods and willows. Along the Missouri the forests range from the water's edge (over natural levees) to a distance of one-eighth to one-half mile from the river above Plattsmouth, and to somewhat greater distances in places farther southward. In addition, they border abandoned river channels and lakes and ponds. In the river, they clothe great sandbars, and sometimes they cover many acres of alluvial deposits on the inner or lower flood plain. The remaining area of the low, wet, first bottomland was occupied by lakes, ponds, marshes, and in places by a luxuriant growth of coarse grasses that often alternated or intermixed with various shrubs.

On the second bottom, forest occupied a very minor portion of the area. Here the supply of soil moisture closely resembles the type now found in prairie elsewhere. Soil development, with rare exceptions, clearly took place under a continuous cover of grass. It is possible that prairie fires had some retarding influence upon tree growth, but it seems more probable that the natural environment of moderate rainfall (23 to 33 inches), dry winds, and long periods of summer drought—all of which characterize the prairie climate—were the determining factors.

Sandbar willow (*Salix interior*) is the first tree or shrub in this area to grow upon sandy or muddy banks of rivers, streams, or lake shores. Its extremely abundant fiberlike roots enable it to maintain a hold on the soil. This small tree, usually about 20 feet high, has a slender trunk that is 2 to 3 inches in diameter. In dense stands, which result from its spreading by long stoloniferous roots to form thickets, it is commonly dwarfed into a shrub only 5 to 6 feet high. Usually there are few other plants in these thick stands, which also are common in drying lakes and ponds. Thickets of willows that cover many acres are frequently found.

One often observes, on the depositing shores of the Missouri, the continuous low zone of sandbar willows that sometimes is many yards in width. Beyond, there is usually a similar stand of tree willows (*Salix amygdaloides* and *S. nigra*). The two species of tree willows are commonly 20 to 40 feet tall and have trunks 8 to 20 inches thick. They thrive on the wet lower banks of streams and on the borders of lakes, ponds, and marshes. Intolerant of shade, they usually form a border to the cottonwood forest, which follows the river all along its course.

Cottonwood is the most typical tree on the banks of great rivers, and often it is the only type on eroding shores where the fringe of willows on lower ground has been swept away. It is the only large tree on the banks of the Missouri and Elkhorn rivers north of central

Fig. 66.—Well-developed flood-plain community on the Big Nemaha River near Falls City, Neb. The white elm in the foreground is 3.5 feet in diameter and 95 feet high.

Nebraska. Cottonwood, which grows rapidly, reaches a height of 70 feet or more, and its trunks are 3 to 6 feet in diameter. Where well-lighted, as on the river margins or in open stands, branching begins low—scarcely higher than the shrubs. In the forest, however, the straight, erect trees are free from branches to a height of 20 to 30 feet.

Scattered over the lower flood plain, far from the river, many cotton-woods of large size occur. Some were 5 to 6 feet or more in diameter, 60 to 85 feet tall, and 70 to 80 years old.

On better-drained alluvial soils farther from the river banks, many other forest trees are common. They are mostly red and white elm and species of ash, boxelder, and hackberry. All of these prefer rich alluvial soils to river sand. Large trees are infrequently found, and dead or dying saplings are common because of poor soil aeration and deep shade. On well-drained, rich bottomlands, however, large specimens are abundant. Black walnut (*Juglans nigra*) is not plentiful, except in the south, where sycamore, (*Platanus occidentalis*) also is a common flood-plain tree.

The willow-cottonwood portion of the flood-plain forest is especially typical of the Platte and Missouri rivers. It extends over low, sandy banks, over sandbars and abandoned channels, and elsewhere. "A later development is shown by the growth of such species as *Acer negundo, Ulmus americana, U. fulva, Fraxinus pennsylvanica,* and *Juglans nigra.* These, in turn, may finally be replaced by *Tilia americana,* the elms often remaining as codominants" (Aikman, 1927). The usual flood-plain species occur on more stable and better-drained soils of the broad flood plain between the bordering wooded hillsides or bluffs (see Fig. 66). The flood-plain community reaches its best development along the larger streams in the southeastern part of Nebraska. Since the shade is denser, fewer of the less-tolerant species are found. The trees are much larger, and in every way the flood-plain forest is better developed. Secondary species are *Fraxinus americana, Prunus virginiana, Gymnocladus dioica, Aesculus glabra, Celtis occidentalis,* and *Platanus occidentalis.* The shrub stage is not as prominent as in the previous stage because of reduced light.

Common shrubs of the flood-plain forest that also extend far beyond its margin and intermingle with coarse grasses follow.

Cornus drummondi Rough-leaf dogwood	*Rhus glabra* Smooth sumac
Amorpha fruticosa Indigo bush	*Prunus americana* Wild plum
Symphoricarpos occidentalis Wolfberry	*Sambucus canadensis* American elder
Symphoricarpos orbiculatus Coralberry	*Ribes missouriense* Wild gooseberry

The following woody vines usually were common, and often abundant: frost grape (*Vitus vulpina*), bittersweet (*Celastrus scandens*), greenbrier

(*Smilax hispida*), poison ivy (*Rhus radicans*), Virginia creeper (*Parthenocissus quinquefolia*), and virgin's bower (*Clematis virginiana*).

It is believed that, along our section of the Missouri River, the flood-plain forest, shrubs, and coarse grasses occupied most of the first bottomland but that most of the second bottomland was covered with prairie. Exceptions, of course, occurred around ponds and lakes and poorly drained land.

The Missouri and Platte rivers have very wide valleys, and although the bordering bluffs and hills furnish protection against the prairie climate, this protection does not extend throughout their width. In fact, some of the best-developed flood-plain forests—with a great variety of trees—occur in the protective area afforded by wooded bluffs and steep hillsides where tributary streams with relatively narrow flood plains join the main rivers. An excellent example occurs where the Big Nemaha joins the Missouri River. It is interesting to note that trees on the upper edge of the protecting slopes are of lower stature than those on the lower slope. Indeed, the canopy of trees that border the streams often is almost level.

The vegetation of the numerous swamps and marshes has been described and illustrated. Bulrushes, cattails, reed, and bur reed, which grow in water 1 to 8 feet deep, were the chief species. Marsh vegetation grew in waterlogged soil where the water level in summer is close to the soil surface. The chief species were grasslike sedges 2.5 to 3 feet high, spike rushes, smartweeds, water hemlock, and many other forbs, such as iris and various mints. The total area occupied by marshes was very large (Weaver, 1960).

Great areas of the flood plains that were intermediate in drainage between the marshes and bluestem prairie supported continuous grassland of prairie cordgrass. It grew in dense stands from thick rhizomes. Other wet-land grasses were rice cutgrass, Virginia wild-rye, reed canary grass, and redtop. Shrubs were often intermixed, as was a large group of tall, coarse herbs.

Transition from this wet-land vegetation is through narrow to wide zones that are dominated by switchgrass, which is a tall, coarse, sod-forming species, and by Canada wild-rye, which also is a coarse grass but of somewhat lesser stature.

The rapid growth and almost complete dominance of big bluestem on the best-drained soils and its possession of the major portion of second bottoms have been shown. Community life in this prairie, the changing structure of the vegetation with the progress of the seasons, and the wonderful productivity of the prairie soils have all been considered (Weaver, 1960).

This study, begun in 1916, endeavored to picture the native vegetation, in its original condition, as a scientific record of the past. After the turn of the century, drainage districts were formed. Lakes and marshes at the margins of the flood plain were filled with soil from the uplands by diverting the water from the hills into them, where the transported soil settled out. Later, powerful machinery was used in clearing away trees and digging deep drainage ditches. Native vegetation throughout the bottomland, except near the river channel, has been almost completely replaced by farm crops.

FORESTS OF EASTERN NEBRASKA

Although general information on this forest area was available at the turn of the century and although detailed studies of forest types near Nebraska City had been made (Pool, Weaver, and Jean, 1918), the distribution and the structure of the several types were made clear by the extensive studies of Aikman (1927). Much of the following data is from this source.

The Missouri River is nearly 0.5 mile wide; its valley varies in width from 0.5 to 1.5 miles where it bounds Nebraska on the northeast. The river is much wider and deeper about 200 miles southward, near the Kansas line, where its valley is 10 to 17 miles in width. Near South Dakota, the flood plain is only about 150 feet below the summit of the first row of bluffs, but in the southeast the bluffs rise to 200 feet or more above the water. Its tributaries and their branches also have developed canyons below the general, windswept prairie level. Here the higher rainfall wets the soil, the large surface of riverwater adds humidity, and bluffs and hills of loess decrease wind movement. Hence a deciduous forest occurs in an otherwise prairie climate.

The widest extent of the forest in southeast Nebraska was 17 to 25 miles, but it was only 5 miles in the vicinity of Omaha, and often it was less than 0.5 mile in the northeast. Although the number of woody species of considerable importance is more than 200 near the center of the deciduous forest of Ohio and Kentucky, the number is only about 80 in southeastern Nebraska. A further decrease of upland woody species, to about 31, occurs where the Missouri River first contacts Nebraska, where the rainfall is only 23 inches.

This forest is composed of three different communities, aside from the valley woodland along the streams, and the communities are named after the most important species.

The community of red oak (*Quercus borealis* var. *maxima*) and linden (*Tilia americana*) is the most mesophytic. Eastward, the oak belongs to the oak forest complex and the linden to that of beech-maple; and it is

Fig. 67.—Interior view of a forest of red oak east of Lincoln, Neb., where the Weeping Water joins the Missouri River. Photo taken in June, 1928.

only because of their topographic relation that they are grouped here. Linden usually occurs on the lower slopes, where water content is higher, and red oak is found in better-drained areas. By the study of 10 well-placed transects across the Missouri Valley, it was found that this community extended as far northward as Ponca, where it occurred only on the lower slopes (see Fig. 67). In southeast Nebraska, it also covered the northern exposures of higher slopes. Red oak attained a height of 75 feet in the southeast, but only 28 feet near Ponca. Likewise, its diameter of trunk decreased from 24 to 8 inches. Linden attained even greater height and girth than red oak, but both dimensions decreased northward, as did those of the oak. Ironwood (*Ostrya virginiana*), a small tree 18 to 30 feet high, is the only species that makes a good growth under the dense forest cover of the red oak and linden, which are larger than those of any other community and produce a thick stand.

Both northward and westward and at the extreme limit of this climax, vegetation covers only the foot of the north slopes. The red oak extends up the Missouri only until the river turns westward. The linden occurs along the foot of the bluffs along the Missouri and Niobrara, and continues westward across the eastern two-thirds of the state. The ironwood ranges still further west, along the Niobrara, and extends into the Black Hills. Where the woods are not very dense in the southeast, Ohio buckeye, Kentucky coffeetree, and redbud (*Cercis canadensis*) may be found, but very few shrubs occur.

The second forest community, of more limited extent, is dominated by black oak (*Quercus velutina*) and shellbark hickory (*Carya ovata*). The trees are about 12 inches in diameter at breast height, and mostly 35 to 50 feet tall. In this less-mesophytic community, especially where the slopes are long and gentle, there is a tendency for the oak to occupy the lower land and the hickory the upper, drier part, but usually they are intermixed. This forest type does not occur beyond Omaha, and its western distribution is somewhat less than that of the red oak–linden community. Other important trees found here are yellow oak (*Quercus muehlenbergii*) and species from the preceding community (see Fig. 68). Shrubs are somewhat more common than in the former community but far less abundant than in the following one, with which they will be described.

The third forest type is dominated by bur oak (*Quercus macrocarpa*) and bitternut hickory (*Carya cordiformis*). It occupies the drier slopes and the tops of hills along the Missouri and also occurs along the larger streams far to the westward (Fig. 69). In the better-protected localities of Nebraska, it attains a height of 85 feet and a diameter of 2.5 feet, but, under extremely adverse conditions, trees 20 to 28 years old may be only 3 to 7 feet high. Scrubby bur oaks are common in ravines near the source of many small streams. Height of the hickory is about 35 feet, but this varies with the habitat from 20 to 70 feet, and the diameter from 4 to 10 inches. A typical community is composed of a somewhat sparse stand of trees, with the oak trees usually greatly outnumbering those of hickory. The fruits of both species are produced in great abundance. The yellow or chestnut-oak is a tree of less importance and is most nearly like the dominants in its habitat requirements. Scattered trees from other communities also occur, and the remainder of this forest is composed of flood-plain species such as ash, elm, and

Fig. 68.—A forest of yellow oak and shellbark hickory in extreme southeastern Nebraska. The shade is dense, and mayapple (*Podophyllum peltatum*) is the most abundant herb on the forest floor.

Fig. 69.—Young bur oak forest near Nebraska City. The trees are about 38 feet high and 10 to 16 inches in diameter.

hackberry that have migrated to the upland. Conversely, bur oak is often found on the flood plains of smaller streams. This forest type is nearly always bordered by a more-or-less-continuous community of various shrubs which separate it from the grassland. Northward, beyond the mouth of the Platte River, the more mesophytic forest communities almost disappear. There is not only a decrease in the number of species and a dwarfing of individuals, but trees are confined to the most favorable sites and the area occupied by shrubs is greatly increased.

It was early pointed out—by Pound and Clements (1900)—that a layer of small trees and shrubs was never well developed in these forests. Rather than in the heart of the forest, the wahoo (*Euonymus atropurpureus*), pawpaw (*Asimina triloba*), chokecherry (*Prunus virginiana*), bladdernut (*Staphylea trifolia*), and rough-leaf dogwood (*Cornus drummondi*) found greater expression in the forest edge, where the forest begins to lose its closed character. Here are found a host of small trees and large shrubs, which constitute immense thickets. The most common examples are thickets of plum, chokecherry, service berry, sumac, hazel, and elder. Other species of almost constant occurrence were gooseberries, Indian currants, and raspberries. Virginia creeper, wild grape, smilax, and poison ivy—all woody climbers—were common in thickets or in the forests' edge, where they climb over shrubs and trees alike and form a dense wall of foliage.

About 40 species of shrubs occur as an understory in the oak-hickory forest in the southeast, according to Aikman (1927). Often they are dwarfed because of the shade, but they flourish in more open places. Some extend far beyond the forest border. In northern Nebraska, the kinds of shrubs are reduced to nearly one-half, but the area occupied by them, with a great decrease in forest growth, is much greater.

The herbaceous flora of the forest is very different from that of the prairie. Forbs of the woodland seldom extend beyond the border of shrubs, and then only in wet years. There is almost no mingling of the two floras.

Aikman (1927) also made extended studies of the environmental factors of both soil and air, and of the correlation of the factors with the type of vegetation. All photographs were taken by the present writer.

EVERGREEN FOREST

In western Nebraska, north of the Niobrara River, the Pine Ridge region extends from Wyoming across Nebraska and into South Dakota. It lies southeast of the Pierre Hills and gumbo plains, and extends eastward to the Sandhill region. The Pine Ridge Escarpment in

Nebraska has an area of about 2,700 square miles. The topography varies from flat to rolling, hilly, or almost mountainous. Much of the area is covered with a dense-to-medium stand of western yellow pine (*Pinus ponderosa*), and scattered trees are common. It is a scenic region, with an altitude of about 4,000 feet (see Fig. 70).

Most of the 18 inches of average annual rainfall occurs in spring. Summer drought is frequent, but the roots of the pine penetrate deeply and spread widely in the sandy, rocky soils. The trees, of medium size, rarely attain a height of more than 60 feet. The trunks are stocky, with a short, clear length, and they seldom exceed a breast-high diameter of 2 feet. This area once yielded trees as large as 4 feet in diameter.

Rocky Mountain or western red cedar (*Juniperus scopulorum*) occurs in some areas, as in northeast Dawes County, where the rough, stony, topography—together with the abundance of lime derived from the underlying chalk rock—forms an ideal soil for this species. The western red cedar is a short, stocky tree that seldom attains a height of more than 20 feet, although the diameter above the ground is 10 to 14 inches. The trunk has no clear length and the lower branches sometimes lie on the surface of the soil.

Species of deciduous trees are common in the draws and canyons. The dominant trees are green ash, American elm, and boxelder—with

Fig. 70.—View in the Pine Ridge country near Chadron in northwestern Nebraska. Photo by W. Tolstead.

small amounts of hackberry. These trees, where they are well developed, are 40 feet or more in height and 6 to 9 inches in diameter. Quaking aspen (*Populus tremuloides*) occurs in some deep canyons, and mountain birch (*Betula fontinalis*) occurs on the banks of some streams, but neither species is abundant. Peach-leaved willow and cottonwood border the banks of small streams. Kellog (1905) and Tolstead (1947) reported on this forest, which also has been described by Weaver (1965).

ROOT SYSTEMS OF TREES

The deciduous forest trees of eastern Nebraska are mostly near the limit of their range, and their grouping according to slope and consequent water relations is clearly defined. Their distribution in ravines and on lower slopes, midslopes, and hilltops was charted in transects at Peru (Pool, Weaver, and Jean, 1916). The well-drained, moist, and sheltered lower slopes are dominated by linden, which intermingles with red oak, and then is replaced by it on midslopes. Here shellbark hickory is usually associated with red oak, but it sometimes occurs in nearly pure stands on slightly drier soils. Bur oak is usually found on the drier upland. This tree distribution was of such regular occurrence that its relation to soil moisture, root habit, and light was thoroughly investigated by Holch in 1931.

Seedlings of each of the preceding species were grown in each of three sites. These were (1) a typical community of linden on a steep, north-facing slope, (2) a gentle southwest slope that was clothed with bur oak, and (3) a slope—then under cultivation—near the crest of the same hill. The sites were within one-fourth mile of each other. In the following discussion of root distribution, the sites will be designated as linden, oak, and prairie (which often clothes such hilltops) stations. The soil at all three stations is very similar, consisting throughout of a deep silt loam known as loess.

During the three years of the experiment, soil moisture was least in the prairie, greatest in the linden forest, and intermediate in the bur oak forest. There was available moisture in the soil at all stations continuously, except in the prairie, where the surface 6 inches were dry for a period of four weeks in 1927 and where the deeper layers were dry for brief intervals. Growth was inversely proportional to the available water content of the soil.

Depth and spread of root systems greatly exceeded the height and spread of tops. The greatest depths attained in the first year were as follows. Bur oak, 5.7 feet in prairie (Fig. 71); shellbark hickory, 2.5 feet; red oak, 2.4 feet; and linden, 1.2 feet. These depths are all from

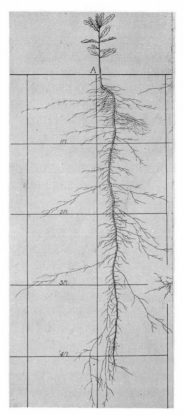

Fig. 71.—Development of roots and tops of bur oak during the first season of growth.

seedlings in prairie. Seedlings in the bur oak forest were only about a foot, or less, in depth. It was necessary to add water to preserve the life of linden seedlings in the prairie. Many seedlings in the linden forest succumbed to the low light intensity. The growth of seedlings was directly correlated with their photosynthetic activity.

The development of bur oak after three growing seasons is shown in Figure 72. Roots of shellbark hickory of similar age in the prairie were about 5 feet deep, with numerous foot-long laterals. Red oak roots were 6 feet deep and much better branched, with long, widely spreading branches. Three-year-old linden seedlings were of similar depth, with widely spread branches (some 3.5 feet long) in the first 2 feet of soil, but these were poorly developed below 2 feet. Thus root development year by year was greatest in bur oak, less in red oak and shellbark hickory, and much less in linden.

Light and soil water have generally been recognized as of major importance throughout the growing season. The form of the root system appears to be correlated with the water supply of the soil. The short taproot and numerous laterals of linden are adapted to the moist but well-aerated slopes just above the flood plain. The well-developed

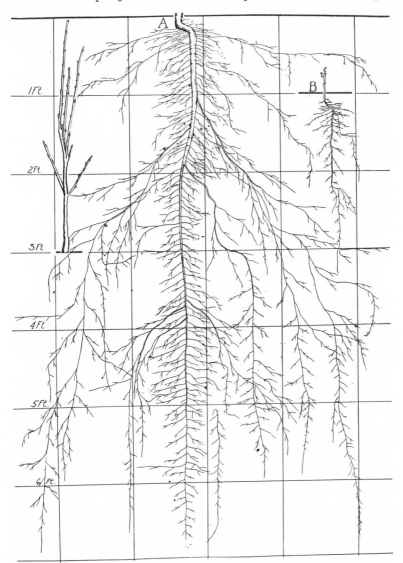

Fig. 72.—Roots and tops of bur oak after 3 years of growth: *A* in the prairie, and *B* in the oak forest. No seedlings survived in the shade of the linden.

taproot of red oak and shellbark hickory fits these species for the drier hillsides above the linden. The vigorous development and extreme length of the taproot of bur oak explain the ability of this species to occupy the drier slopes and hilltops and to endure the drought most successfully. The bur oak seedlings are indeed well adapted to upland sites. The taproot is usually about 9 inches deep before the first leaves are unfolded. In contrast, those of the linden are only 1 to 1.5 inches in length.

Depth and spread of root systems greatly exceeded the height and spread of tops. Even after 3 years of growth in full sunshine, the greatest height (only 3 feet) was attained by the bur oak. The preceding data are from Holch (1931).

An ecotone between grassland and deciduous forest along the Missouri River occurs in southeastern Nebraska. The occupation of grassland by forest is often attained only as a result of changes in the habitat that are brought about by shrubs. In some places, however, bur oak, the most xeric forest tree, has invaded the grassland without the the aid of shrubs. Here competition for water between the grasses and the invading trees is great.

In southeastern Nebraska, near Nehawka, the root systems of several mature bur oak trees were examined, and that of one tree was completely excavated (Weaver and Kramer, 1932). Here the mean annual precipitation is 31 inches, humidity is relatively low, and evaporation and wind movement are relatively high. The silt loam soils are deep, but during drought they often contain only a small amount of water that is available for growth. Trees 50 to 65 years old, which grew 10 to 40 feet apart in a pure stand on the slope of a south-facing hillside, were examined.

The roots of a mature tree—37.5 feet tall, 14 inches in basal diameter, and 65 years old—were selected and the root system was excavated. The taproot, tapering rapidly, gave rise to about 30 large main branches, most of which arose in the first 2 feet of soil. Neither the taproot nor its branches extended deeper than 15 feet, and most of the main branches were 1 to 7 inches in diameter. They extended outward 20 to 60 feet before turning downward. Branching was profuse and a very large volume of soil was occupied. Many branches of the main roots grew almost vertically downward for 8 to 15 feet. Each, with its branches, more or less resembled the taproot system of an oak sapling. Other branches extended vertically upward and filled the surface soil with a network of absorbing rootlets. Still other rope-like roots, a half-inch or less in diameter, extended many feet without much change in thickness. A cordlike type, only 3 to 5 millimeters thick, also was abundant. A third type clothed the widely extending skeletal framework. They consisted of fine, much-branched rootlets, and furnished the bulk of the absorbing system. Mycorrhizal mats were abundant.

It was ascertained that the weight of the roots equaled that of the tops, and the volume of the roots was only one-tenth less than that of the parts above ground. That the adaptation of a species to its habitat is largely a matter of root development is a viewpoint that is strongly supported by rapidly accumulating evidence (Weaver and Kramer, 1932).

An experimental study of the development of the roots and shoots of eight species of deciduous forest-tree seedlings and saplings was made by Biswell (1935). The root systems of seedling honey locust (*Gleditsia triacanthos*) were 1.5 times as deep as the height of the tops, and those of shellbark hickory (*Carya ovata*) were 10 times as deep. Other species were intermediate.

Honey locust has a generalized root system that is readily modified by environment. In upland soil, taproots of saplings penetrated to 5 feet, but on a flood plain their depth was only 2 feet—although the laterals extended outward 10 to 17 feet. Boxelder (*Acer negundo*) has a very plastic root system which developed in a manner similar to that of the honey locust.

Young delicious apple trees were grown in clay loam soil at Lincoln, Nebraska, and in silt loam soil (loess) at Union, about 30 miles eastward. The development of roots and tops was studied during the first three years after transplanting selected two-year-old nursery stock in the orchards. Large numbers of transplanted trees were examined after each of three years of growth (Yocum, 1937).

The development of the root systems was extremely rapid. Roots reached a maximum depth of 8.8 feet and a lateral spread of 12 feet the first year, and 14.8 and 21.2 feet the second year. During the third year, the maximum lateral spread was 29.4 feet and the maximum depth was 17 feet.

After three years under clean cultivation, the trees had an average lateral root spread of 23.6 and 19.2 feet at Lincoln and Union and an average root depth of 9.4 and 14.7 feet, respectively. The root systems were very plastic and were greatly modified under competition with corn—and under both a straw mulch and a sod mulch. The roots of 73 trees were examined.

DEATH OF TREES DURING THE GREAT DROUGHT

Several years of decreasing precipitation initiated the seven years of drought. The 12-month period following June, 1933, was the

Fig. 73.—Loss of elms by drought along the Weeping Water River in eastern Nebraska. The elms were growing on the flood plain and were protected from drying south winds by a steep bluff. Photo, taken in May, 1944, shows some recovery.

driest-weather period ever recorded in several midwestern states. This intensely dry period, as well as those in several of the following years, was accompanied by record-breaking temperatures, extremely low humidities, and exceptionally high rates of evaporation. High winds, swarms of grasshoppers, and great duststorms also prevailed.

Where woodland extended far westward along the Missouri and its tributaries, the loss of trees was extremely heavy. This took place despite the fact that in this prairie area only the most drought-resistant species occurred. The following description is a summary of several years of study on the injury and death—or recovery—of trees in prairie climate, which is a paper of 38 pages and 62 figures (Albertson and Weaver, 1945).

The early effects of the drought were very impressive and widely reported. The literature revealed that, in Minnesota, 40 percent of all the trees in shelter belts—mostly boxelder, willow, green ash, silver maple, and cottonwood—were considered dead or dying in 1934. In North Dakota, practically all maturing cottonwoods, willows, and box-elders were killed, except where the water table was continuously high (the boxelders often recovered by root sprouts). Great damage to trees also occurred in Montana. Losses of coniferous trees in Iowa in the spring of 1934 ranged from 5 to 39 percent. Trees in central Kansas showed 41 percent death or injury, and those westward showed 55 percent, in 1936. In central Oklahoma, 20 to 50 percent of the trees died, and much higher losses—35 to 79 percent—were reported in western Oklahoma by 1937 (see Figs. 73, 74). Reconnaissance of tree plantations late in 1934 showed a 43 percent survival of standing trees in North Dakota, 32 percent in South Dakota, 18 percent in Nebraska, and 28 percent in the Oklahoma–northern Texas area. As the drought progressed, earlier losses were greatly increased.

Pre-drought data on mean annual precipitation, average weekly evaporation, and the water content of the soil in eastern Nebraska, north-central Kansas, and eastern Colorado are presented. They clearly show the semiarid environment west of the Missouri River and why this becomes even more inimical to growth of trees farther westward. A period of dry years preceded the severe drought. Precipitation was lowest in 1934 and 1936, but, because of previous desiccation of

Fig. 74.—Shallow ravine near Alma in south-central Nebraska in which American elms were all dead in August, 1939.

Fig. 75.—Interior of a woodland of shellbark hickory (*Carya ovata*) near Nehawka in eastern Nebraska. Before the drought this forest of 35- to 45-year-old trees was in good condition, but in 1943 scarcely an uninjured tree was found.

soil and vegetation, 1939 also was one of the worst drought years. The extremely low precipitation during these years and the weekly amounts of available soil moisture (which very frequently were nil to 6 feet in depth during the drought) were ascertained. The water table in many ravines and lowland terraces fell 3 to 4 feet or more over much of the Midwest—and trees on an upland root far above the normal water table. Evaporation was extremely high, and sometimes was one-third or more greater than during a pre-drought year. Increases of 6° to

15° F. in the average weekly temperatures were common, and average weekly maximum temperatures of 101° to 109° F. sometimes occurred.

The chief cause of injury was a lack of sufficient available water. This was due to low precipitation but it was accentuated by one or more of several causes: competition for water by grasses, decreased rate of water infiltration and rapid runoff, drying up of streams and springs, and a rapid fall of the water table in ravines and lowland terraces. Lack of an adequate water supply also resulted from low humidity, high evaporation, desiccating winds, and the inability of trees to accommodate their root systems to the rapidly changing environment. Unrestricted grazing was a common cause of excessive mortality in timber claims and windbreaks (see Figs. 75, 76).

Experimental data for the harmful effects of competition with grass both on the roots and shoots of trees are presented. Root distribution of the same tree species in different types of soil, and the general relationship between root extent and distribution in different sites to drought resistance, are pointed out.

Fig. 76.—Remnants of a timber claim on a level lowland near Broken Bow, Neb. An abundance of buffalo bur (*Solanum rostratum*) and other weeds indicate much grazing and trampling. The green ash that had not died showed life only in the lower portions of their crowns.

Injury and death of woody plants due to the effects of drought were often the results of continuous adverse conditions over an extended period of time. But the death of trees and shrubs on flood plains and terraces sometimes occurred in a relatively short time if the water table lowered rapidly. The effects of early drought were wilting, discoloring, withering, or shedding of foliage. An early outward sign of repeated yearly drought among deciduous trees was great reduction in the size and number of leaves and the defoliation of the outer portions of the crown. Also, great injury was often caused by partial or total—and sometimes repeated—defoliation by grasshoppers and web worms and by the leaf-eating larvae of other insects. Such attacks usually occurred during the years of great drought.

Exposure of branches with reduced foliage to high insolation, to great heat, and to low humidity was a common cause of injury. Desiccation resulted in the death of the smaller branches, which permitted the entrance of wood-borers, other insect larvae, and fungi. Desiccation and wood-borers caused the death of the branches to proceed rapidly downward; often the entire tree succumbed.

Losses in the most xeric places, where trees grew naturally, were high. In dry ravines, American elm (*Ulmus americana*) losses were 70 percent and hackberry (*Celtis occidentalis*) 36 percent; on bluffs, American elm losses were 56 percent, hackberry 28 percent, and green ash (*Fraxinus pennsylvanica lanceolata*) 33 percent. Losses of red cedar (*Juniperus virginiana*) usually were less than those of the preceding species. Average percentage losses of trees along continuously flowing and intermittent streams were American elm, 5 and 62 percent; hackberry, 5 and 75 percent; and cottonwood (*Populus deltoides*), 6 and 59 percent. Loss of willows (*Salix* spp.) along intermittent streams was 70 percent. Death of trees growing near running springs was rare, but near springs that almost failed to flow in the drought a mortality of 55 percent for cottonwoods and 89 percent for willows was recorded.

Destruction in long-established timber claims was high. Severe losses occurred during dry periods previous to 1933, and the great drought almost annihilated the survivors. Although 45 percent of osage orange and 86 percent of green ash were dead by 1935, all had succumbed by 1939. Even in tree claims on lowlands early in the drought, black locust (*Robinia pseudoacacia*) had lost 75 percent and cottonwood had lost 14 percent.

Losses of trees in windbreaks were nearly always heavy. Dust-storms contributed greatly to this loss through partial burial of the trees in great drifts of soil, sometimes 4 to 8 feet deep. Losses of green ash, American elm, hackberry, and other deciduous trees frequently were 80 to 90 percent, and sometimes higher. In some places, trees of red cedar alone survived. Of 3,200 trees in a hedgerow of osage orange (*Maclura pomifera*), all but 15 succumbed. Such high losses were typical.

Average losses of trees in Nebraska and Kansas probably were 50 to 60 percent by death, and an additional 20 to 25 percent suffered severe to moderate injury. Red cedar, hackberry, and bur oak endured drought especially well; silver maple, boxelder, and cottonwood seemed less well adapted to endure long-continued drought.

An extensive study was made on the effect of a changing environment on trees as recorded by the annual increment of radial growth.

Trees that retained some life at the close of the drought usually remained alive unless infestation by wood-borers was so complete or the trees so nearly dead that they were unable to resume growth. Recovery was shown principally, and commonly, by renewed growth locally within the crown. In dry sites even after 3 or 4 years of good precipitation, leafy branches were sometimes few and foliage was sparse. But where drought had been less severe, the foliage of the renewed portions of the crown was unusually dense. Often, where moisture was plentiful, the dead branches in the tops of the crown were soon obscured by new ones. Sprouts that developed from the bases of certain trees grew rapidly. Dead trees were partly replaced by seedlings, but only where the trees grew naturally. In this manner, red cedars continued to replace their losses throughout the drought. Seedlings were not found in timber claims, windbreaks, or hedgerows in mixed prairie, and only rarely in unpastured places eastward to the Missouri River.

In the pine forests of the northwest, the deeply rooted yellow pine suffered only small losses, but the broad-leaf species in ravines and along streams—with roots extended into moist soil or to water—suffered heavy losses with the lowering of the water table.

VIII.

Beginning of Seven Years of Drought

Measurement of the environmental factors in prairie, in connection with various grassland researches, had been completed by 1928—except for one year—for the thirteenth consecutive growing season. Precipitation had been somewhat deficient and periods of drought occurred in 1931 and 1933, when moisture was not available to the roots of plants in the surface 12 inches of soil. This, however, had little harmful effect upon the deeply rooted prairie vegetation.

DROUGHT IN TRUE PRAIRIE

After an unusually warm winter with light snowfall, the spring of 1934 began very dry. From March until June, the total rainfall was only 0.9 inch. This introduced the summer of 1934, when drought was the greatest ever recorded in true prairie. This event offered an exceptional opportunity to study the response of the native plants to extremely adverse water relations (Weaver, Stoddart, and Noll, 1935). Moreover, prairies used for the drought studies were the same as those that had been employed in the investigation for "The Prairie" (Weaver and Fitzpatrick, 1934). Each prairie was known intimately; hence any changes were readily and clearly observed.

Studies of the water content of the soil clearly showed that the drought came on gradually during a period of three or four years. Soil moisture on uplands was slowly depleted, and by July 30, 1934, no water was available to a depth of 4 feet for plant growth (see Fig. 77). Arid conditions were accentuated by high winds, of an average mean daily velocity of 12 miles per hour. Frequently, in spring, the wind carried great quantities of dust which lodged among the grasses and readily rose again into the air. During July of this abnormally hot growing season, the average weekly maximum daily temperature in the prairie varied from 98° to 111° F. The average weekly maximum humidities varied from 15 to 22 percent. During certain afternoons, humidity was only 3 to 5 percent. From hilltops, drought swept down the slopes into mesic and hydric ravines. The wilting, drying, and death of plants was not due only to high temperatures and low humidities

146

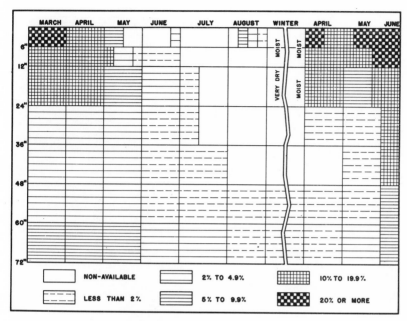

Fig. 77.—Available water content of prairie soil near Lincoln, Neb., at several depths (to 6 feet), from March, 1934, to June, 1935.

but primarily to the low water content of the soil, since plants in watered areas thrived. Drought continued during the winter and was not ameliorated until the following spring (see Fig. 77).

A close relationship was found between the root depth of most prairie grasses and their resistance to drying. Among forbs also, resistance to drought was closely correlated with root extent. Species with root systems that penetrated 8 to 20 feet into moist soil were little affected. This is in accord with the investigations of Nedrow (1937), who found a direct relationship between the amount of growth of tops of dominant prairie grasses and the depth at which water was available for absorption. When water was available only at a depth of 3 or 4 feet, some development of tops still occurred (Fig. 78). Certain forbs grew normally when water was available only at depths below 3 to 5 feet.

The blossoms of deeply rooted plants marked the vernal and estival aspects. Flowering often began two to three weeks early and was of shorter duration than normal. Species of Viola, Oxalis, Fragaria, and Vicia developed only poorly. Not infrequently, the flowers withered and dried, and the production of viable seed was almost nil. Most of these plants were rather shallowly rooted. *Poa pratensis, Koeleria cristata, Antennaria neglecta,* and other shallowly rooted species dried in May.

Fig. 78.—New growth of *Poa pratensis* between June and September, 1934. Each container holds a core of bluegrass sod that is 10 inches in diameter and 2.5 feet deep. The cores were removed intact from the prairie at a time when the soil was dry throughout and the plants appeared to be dead. The cores were watered at depths of 4, 15, and 24 inches, respectively.

The drying bluegrass gave the prairie its early, dead appearance, and the abundance of this invader could be estimated readily by the color of the landscape. *Andropogon scoparius* in the dry upland soils withered early in June; *A. gerardi*, because of its deeper root system, persisted for a longer time. *Stipa spartea* and *Bouteloua gracilis* were more resistant; they rolled their leaves and assumed a condition of drought dormancy.

The habitat dominated by *Andropogon scoparius* was one of more favorable water relationships, compared with the habitat of the needlegrass–blue grama community. This grass dominates on the steep hillsides that are characterized by bunch grasses. It forms distinct bunches in dry places, but a nearly continuous sod-mat in places that are less dry. On gentle slopes, it is also often found in scattered bunches among the short grasses.

Andropogon gerardi dominates the lower slopes and ravines where runoff water and blown-in snow augment the precipitation. Here it usually composed 75 percent of the vegetation. It was accompanied by *Panicum virgatum* and *Sorghastrum nutans* from true prairie, but these tall grasses do not attain the stature of the same species eastward.

EFFECT OF THE DROUGHT OF 1934

The study in true prairie was continued in 1935 in order to ascertain the results of the drought of 1934. It was extended to include the mixed prairies of west-central Kansas where an even more severe drought had prevailed for the two preceding years (Weaver and Albertson, 1936). Drought in the true prairie will be described first.

Figure 77 shows that available soil moisture at the Lincoln station during the winter of 1934/1935 occurred only below 4 feet. By June, 1935, subsoil moisture had been replenished to a depth of 6 feet. At the 160-acre Belmont prairie at Lincoln, which was representative of numerous prairies of eastern Nebraska, the little bluestem type alternated with needlegrass on the hilltops and xeric slopes, and with big bluestem in the ravines. Losses from drought varied greatly with slope and exposure. The prairie was mowed once annually, which was the usual practice. The major losses were of the nature of holes or openings in the prairie carpet, a condition that was greatly emphasized because of the unusual development of the foliage cover in 1935. Superficially, the prairie appeared normal; to appreciate the losses, one had to penetrate the foliage and study the basal cover. Each part of the prairie bore its toll of dead crowns of little bluestem or prairie dropseed, dead rhizomes of big bluestem and Indian grass (often exposed by the erosion of wind and water), pale stem-bases of needlegrass and blue-grass, etc. (see Fig. 79). Many such areas remained bare throughout the entire growing season; others were marked by dense patches of annuals or short-lived perennials, such as *Lepidium virginicum* and *Festuca octoflora*.

From a large series of permanent chart quadrats established before the drought, a loss of 36 percent of the basal area between 1930 and 1935 was ascertained. Where drought was most severe, the open spaces were more plentiful and formed a very irregular network of unoccupied soil surface. Thus openings of moderate size graded imperceptibly into larger areas of a square meter or more, which also were destitute of living plants. Losses on many uplands near Lincoln were only 10 to 25 percent and a good matrix of native grasses remained. In other areas, more than half of the bluestems and other grasses had been killed, and forbs—except deeply rooted ones—had suffered similar losses. Often, only the more xeric grasses—*Bouteloua gracilis, Stipa spartea,* and remnants of *B. curtipendula*—remained.

Fig. 79.—Bare areas in a little bluestem prairie in 1935 as a result of the drought of 1934.

Great destruction to species of the understory with the opening of the taller grass cover gave the prairie its distinctly open appearance—such lower-stature species as *Panicum scribnerianum, Eragrostis pectinacea, Antennaria neglecta,* and *Senecio plattensis. Agropyron smithii,* western wheatgrass, which formerly occurred very sparingly, made notable increases not only along roadsides and in pastures, where it became extremely common, but it also invaded the drought-stricken prairies.

The Belmont prairie was only one in a group of eight large grassland areas that were thoroughly examined in eastern Nebraska. Several well-known prairies in southwestern Iowa also were reexamined, as well as four prairies in north-central Kansas on the western edge of true prairie. Drought had been least severe in Iowa and most destructive in north-central Kansas, where, in many places, 75 percent of the former vegetation had died and where often—on flat hilltops— only big bluestem was left. In other places, even on lowlands, losses of this deeply rooted species were great. Invasions of wheatgrass increased and the spread of buffalo grass and blue grama became locally common.

The 30 or more prairies examined in Iowa, Nebraska, and east-central Kansas had been fully studied before the drought, so that changes were readily determined. Those on the deep loess soil of south-western Iowa had not been harmed. Other prairies had suffered losses of from 20 to 50 percent on thinner soil of exposed ridges to 80 to 95 percent on nearly level areas farther westward. Great destruction also occurred even on low ground, which in eastern Nebraska sometimes resulted in an entire change of plant populations.

All of the native grasses suffered some loss, but death was greater among those with relatively short roots, such as little bluestem, June-grass, needlegrass, and the invading Kentucky bluegrass. This often occurred generally, but it was especially common where they occupied the drier soils. Prairie dropseed also sustained losses, which, like those of Indian grass, often amounted to 80 percent or more. Big bluestem, because of its deeper root system, usually was injured least. Losses of the interstitial low grasses and forbs often were nearly complete after the protecting cover of tall grasses had dried.

Invasion by the drought-resisting and rapidly spreading western wheatgrass was rapid and locally complete. It occurred widely. Buffalo grass and six-weeks fescue were other native grasses that increased greatly, especially westward. Some of the native forbs that spread most widely, after the death of the dominants decreased competition, were *Aster ericoides*, *Erigeron ramosus*, *Silene antirrhina*, and *Specularia perfoliata*. Certain ruderals, normally not found in prairie, had been widely distributed. The most conspicuous were *Lepidium virginicum*, *Leptilon canadense*, *Bromus secalinus*, and *Tragopogon pratensis*.

MIXED PRAIRIE BEFORE THE DROUGHT

Following these initial reports on the beginning of drought, the study was extended to include the central portion of the Great Plains, where work was continued jointly with Dr. F. W. Albertson, Professor of Botany in the Fort Hays Kansas State College, at Hays, in west-central Kansas. Dr. Albertson, a native of Kansas, was well acquainted with the flora of the Great Plains even before his doctoral thesis, "Ecology of Mixed Prairie in West Central Kansas," was begun (Albertson, 1937). The main area for detailed study was a large tract of native mixed prairie (750 acres), formerly a small part of a military reservation, which had been fenced and only lightly pastured since 1900. Studies of the vegetation were made almost continuously through a period of four years. Numerous other large areas of mixed prairie also were studied in order to obtain a better background for the intensive work.

List quadrats, and especially permanent chart quadrats, were used extensively, and transects were employed. Many bisects were made. Trenches were excavated in all types of soil, and often through outcrops of limestone rock in order to ascertain the relationships of underground plant parts to rock crevices and soil. Community relationships both above and below ground were fully examined, which involved detailed study of the more important species. This was an ideal background for an understanding of the drought which swept the central Great Plains and prairie during a period of seven years.

A brief statement of the types of pre-drought vegetation following several years of good rainfall is a necessary background. Three types of vegetation were common: the short-grass community, the little bluestem community, and the big bluestem community. The appearance of the short-grass vegetation was that of a well-grazed meadow, except for the presence of certain taller grasses and forbs. These were typically scattered throughout, or locally, where conditions were more favorable, they became the conspicuous features of the landscape. The two short grasses (*Buchloe dactyloides* and *Bouteloua gracilis*) alone furnish fully 80 percent of the vegetation. The foliage of the grasses was only 3 to 5 inches high and of a dark-green color when moisture was plentiful. With the advent of drought, the grasses became bleached to various shades of yellowish green, and finally to a light gray. After the drought was broken, they soon resumed their original green color. These conditions occurred almost every season. The cured condition of dormancy is the prevalent one during years of drought (Fig. 80).

The density of the vegetation varied from that of the usual closed-mat type, in which 70 to 90 percent of the soil was covered, to the open-mat type of poorer soils, where only 20 to 30 percent was clothed during a season of normal precipitation. The foliage cover of the short grasses during dry years scarcely exceeded the area of the plant bases, but it was somewhat greater under conditions that were favorable for growth. Approximately 95 percent of the prairie species are long-lived perennials, a condition that prevails in the other communities as well. Taller grasses, if they were thickly grouped, formed a distinct upper layer, but these bunch grasses usually were not abundant. The most important were *Aristida purpurea*, *Andropogon scoparius*, *Bouteloua curtipendula*, and *Sitanion elymoides*.

The little bluestem type occupied the hillsides and extended across ravines and over the brows of the hills—and far beyond, where the slopes continued. It gave way, more or less abruptly, to the mixed prairie type (mostly short grass) on the level uplands. Accompanying species were big bluestem, switchgrass, and side-oats grama. Distribution of this type was largely controlled by the shallow soil, which was about 2 feet deep above the crevices and pockets of the underlying

Fig. 80.—Short-grass country—blue grama and buffalo grass—in eastern Colorado before the great drought.

limestone. On the nearly level uplands, where a mature soil profile has developed, it is replaced by the short grasses.

At Hays, the short-grass–mid-grass type of mixed prairie was characterized by a predominance of buffalo grass and by blue grama, about one-third as great in abundance. Together they composed about 80 percent of this type. The chief mid grasses were wire grass (*Aristida purpurea*) and little bluestem. Another mid grass was western wheatgrass, which usually forms streaks or patches, but, like sand dropseed (*Sporobulus cryptandrus*), it was often scattered thinly in the short-grass sod.

On lowlands where deep alluvial soil is moistened by run-in water, little bluestem is replaced by a tall-grass community. Big bluestem is the chief dominant, often composing 80 percent of the vegetation, but side-oats grama, western wheatgrass, tall dropseed (*Sporobolus asper*), and switchgrass also were common.

EXTREME DROUGHT IN MIXED PRAIRIE

Severity of drought in mixed prairie was studied intensively at Hays, Kansas, and from Nebraska to the Oklahoma border in west-central Kansas. These studies of the environmental factors and the structure of the mixed prairie had begun before the advent of the great drought and they continued until the end of 1935 (Albertson, 1937). At Hays, a tract of 750 acres of unbroken prairie was used. This prairie covered a rolling topography that was adjacent to the valley of a branch of the Smoky Hill River. The elevation was about 2,100 feet, and the level hilltops rise about 200 feet from the valley floor. The three general types of vegetation, with varying degrees of intermixtures, were common.

The drought years of 1933, 1934, and 1935, when the mean annual precipitation was only 15.4 inches, were preceded by six years with an average precipitation of 4.6 inches above the normal (22.8 inches). During the drought the water content of the soil was exhausted to—or almost to—the depth of root penetration of the dominant grasses (3.5 to 6 feet). This was ascertained by weekly soil samplings to a depth of 2 feet in the little bluestem type and to 5 feet in the other communities. Moreover, depth of root penetration, branching, etc., were ascertained in all habitats by a series of trenches dug to depths of 3 to 13 feet and totaling 400 feet in length. The great soil drought occurred at a time when the average day temperatures were 86° to 97° F. and the relative humidity was very low—often only 11 percent, or less, in the afternoons. These severe conditions were aggravated by high winds and often by duststorms.

Andropogon scoparius suffered losses of 90 to 100 percent where it was intermixed with the short grasses. In its own community, losses of 50 to 87 percent occurred where it was ungrazed, and losses of 66 to 96 percent occurred under moderate grazing. Moderate grazing that preceded the drought reduced the density of the cover and enabled the vegetation to survive longer on the remaining soil moisture. Side-oats grama and big bluestem, both intermixed with but more deeply rooted than little bluestem, at first suffered losses. With temporary conditions that were favorable to growth, they recovered and partially replaced little bluestem.

The closed type of buffalo grass and blue grama showed relatively small losses when they were protected from grazing, although certain native forbs entirely disappeared. Losses in open, ungrazed short grass were 70 to 80 percent. In both types, stolons of buffalo grass rapidly reclaimed bare areas when moisture was available for growth (see Fig. 81). Gains of 30 to 111 percent in basal cover over that during the spring of 1935 were recorded in the fall of that year. There was a marked tendency of this grass to reclaim areas that had been bared by the death of the mid grasses. At Phillipsburg—between Holdrege, Nebraska, and Hays, Kansas—an area of mixed prairie was converted

Fig. 81.—Stolons produced by a small mat (center) of buffalo grass at Lincoln, Neb., during a single summer with good rainfall.

into pure short grass by the death of the bluestems and other grasses. Other prairies were greatly damaged, and often were buried by dust from cultivated fields and converted into waste areas that were populated by weeds (Fig. 82).

Ungrazed prairies in south-central Kansas lost 60 percent of their basal cover; various types that were moderately grazed lost 36 percent; and heavily grazed prairies lost 74 percent. Losses, in the same sequence, in north-central Kansas were 50, 54, and 91 percent. Conditions were most severe in the west-central part, where ungrazed prairies lost 85 percent, moderately grazed areas lost 72 percent, and heavily grazed areas lost 91 percent. The cause of this destruction was deficient soil moisture, which was coupled with extremely high temperatures and low humidities and supplemented by wind and burial by dust.

Changes in the populations of grasses and forbs were pronounced. Certain grasses that formerly were found regularly in midwestern Kansas were found only occasionally; examples are *Koeleria cristata, Poa arida, Aristida purpurea,* and *Festuca octoflora.* The following forbs were either partially or completely eliminated from mixed prairie: *Achillea lanulosa, Ambrosia psilostachya, Astragalus shortianus, Oxytropis lambertii,*

Fig. 82.—A range near Winona, Kan., with only 5 percent cover, where the only remaining grass is blue grama (white patches). The soil has been bared by drought, except for Russian thistle and other annual weeds. Photo was taken in 1939.

and 15 others. Conversely, the decrease in the ground cover occasioned by the death of the grasses and forbs resulted in an increase in the numbers of certain native forbs of greater drought resistance. Among this group of 14 species, *Aster ericoides, Diaperia prolifera, Gutierrezia sarothrae,* and *Malvastrum coccineum* are examples.

Seeds of ruderals had been spread widely by the wind over the whole Midwest. Ecesis in the bared areas was prompt upon the advent of rain. Consequently, most prairies and pastures were literally covered with seedlings of species of Chenopodium and Amaranthus, *Eragrostis major, Panicum capillare, Setaria viridis, Solanum rostratum, Lepidium virginicum, Salsola pestifer, Bromus tectorum* and *Hordeum pusillum.* So abundant were the weeds that the prairies often appeared more like abandoned fields than grassland. The tallest weeds had rather completely covered the soil in 1935, but the summer drought had killed most of them before they were fully grown.

The large number of chart quadrats that show the changes in the several types of vegetation from year to year (Weaver and Albertson, 1936) are of special value.

COMPETITION OF WHEATGRASS IN RELICT TRUE PRAIRIE

Agropyron smithii is a common sod-forming, perennial, forage grass of midwestern prairies. It is so successful a competitor for the meager supply of soil moisture that it often causes the death of the more mesic grasses and forbs of the true prairie. It renews its growth in early spring, produces abundant foliage which normally reaches a height of 1.5 to 2 feet in June, and is overtopped by flower stalks that are 1.5 to 2 feet taller. Seed is produced in abundance, and migration is rapid by means of long, slender, much-branched rhizomes. Formerly occurring sparingly in the eastern portions of Nebraska and Kansas, western wheatgrass spread rapidly and widely following the great deterioration of grassland due to drought.

Its early, luxuriant growth, when water was available, resulted in greatly reducing the amount of soil moisture for use by other species, most of which began development four or more weeks later (see Fig. 83.) Lack of much debris under western wheatgrass permitted the rain to loosen the surface soil and to roil the water that entered it. This resulted in decreased infiltration and greater runoff than on soil that was covered with bluestems.

The rate of entry of water was determined for soil covered with bluestem or other pre-drought native grasses and for the same soil clothed with western wheatgrass. In these experiments, steel cylinders 1 square foot in cross-sectional area and 4 inches long were used. The cylindrical wall was only 2 millimeters thick and the steel was sharpened; hence, after it was oiled it could easily be forced vertically into the soil, without disturbance to soil structure, to a depth of 3.75 inches. Water was then added, as rapidly as it could be absorbed, from a sprinkling can with small perforations, until a total of 1 gallon had been applied. The time for the infiltration of the water was recorded, and was consistently much greater in western wheatgrass. The average time was 6.4 minutes in a stand of prairie grass and 15.3 minutes where wheatgrass clothed the soil. Thus the rate of infiltration was 2.4 times as rapid in normal prairie as in wheatgrass. These data are in accord with later studies by Robertson (1939), who ascertained that—of the 12 types of vegetation he measured—soil covered with big bluestem absorbed water most readily and soil occupied by western wheatgrass absorbed water least rapidly.

The amount of water transpired from a normal stand of western wheatgrass or evaporated from the soil it covered was more than twice as great from March 25 to May 20 as that from little bluestem (see Fig. 83).

Normal root depth in moist soil is about 8 feet, but during the drought the depth corresponded with the depth of moist soil, which

Fig. 83.—Development of transplanted sods of little bluestem (top left) and of western wheatgrass (top right) on April 25. The bluestem averaged 1 inch in height, and only half of the sod was green; the wheatgrass was 7 inches tall and in the fourth-leaf stage. The lower section shows the condition of the plants on May 20.

was about 2 to 2.5 feet. A few feet distant, higher water content and much deeper penetration of both water and the roots of other grasses were recorded.

Competition for water resulted in great dwarfing and often in wilting and the death of most other prairie grasses and forbs. Numbers of species and numbers of stems of perennial forbs were greatly decreased after western wheatgrass once became thoroughly established. In prairies on silt loam soil and under similar precipitation, the number of perennial species of forbs was only 56 percent as great in wheatgrass, and the number of stems was 20 percent of that in uninvaded areas (Weaver, 1942).

The large area of drought-damaged true prairie and of native pasture that was dominated by western wheatgrass in 1941, together with the harmful effects of the successful competition for water by this grass with species of greater forage value, presented a problem of much scientific interest and of great economic importance.

IX.

The Great Drought

The drought of 1933 was only the beginning of a long period of deficient rainfall. The late summer of 1934 was a period of great drought, and the summer of 1936 was the hottest and driest ever recorded in eastern Nebraska (Weaver and Albertson, 1936). Since one cannot appreciate this condition by merely reading the fact, the experience of Robertson (1939) will be helpful.

> Four days spent on this prairie in the middle of July, 1936, served to impress one with the severity of physical and biotic factors so extreme that even native vegetation could not endure them uninjured. Daily maximum temperatures ranging from 104° to 111° F. accompanied by relative humidities of 19 to 24 percent, continued strong winds, glaring sunlight, and subnormal precipitation combined to make the grasses crackle underfoot like wheat stubble. Only the very deeply rooted false boneset (*Kuhnia eupatorioides*) and blazing star appeared unhampered by drought, and they were borne down and partly eaten by hordes of grasshoppers.... Buffalo grass foliage, including stolons as much as 16 inches long, was bleached and apparently lifeless. Bunches of blue grama, some a foot in diameter, seemed entirely dead.

Drought years continued until 1941. Stations in the area of great drought were visited every two weeks despite the weather, and soil samples for determining water content to depths of 5 to 6 feet were secured (see Fig. 84). The study was continuous from 1933 to 1941 (Weaver and Albertson, 1939, 1940). During the following ten years the return of prairie to its former climax condition was recorded year by year. The several publications on the drought and recovery furnish the most complete record of these eventful happenings.

REPLACEMENT OF TRUE PRAIRIE BY MIXED PRAIRIE

The replacement of true prairie by mixed prairie occurred as a result of drought in an area 100 to 150 miles in width. It was due to a dry climatic cycle and was accomplished within a period of seven years. The scene of this action was the broad area on the western edge of true prairie—in central Kansas, eastern Nebraska, and eastern South Dakota. It is distinctly east of the mixed prairie in the central and

160

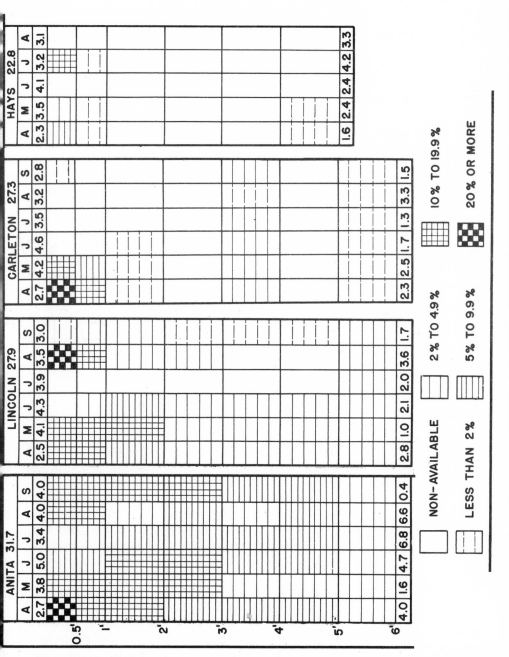

Fig. 84.—Available soil moisture, to a depth of 6 feet, at four stations in 1940. Mean annual precipitation at each station follows the name, mean monthly precipitation is given below the letters that indicate the months, and current monthly rainfall is given at the foot of each column.

western portions of these states. Although the phenomenon was general over hundreds of square miles of grassland, it was studied in detail from year to year in only a number of widely separated places. Three of these places have been selected to illustrate this remarkable transition in grassland populations and structure and the manner in which it occurred—as well as the time relations and the casual factors.

The three prairies selected for illustration had been studied at two periods just before the drought and intensively—summer after summer —following 1934. They are located 65 miles southwest of Lincoln, near Carleton, Nebraska; 30 miles northwest of Carleton, near Clay Center, Nebraska; and 45 miles southwest of Carleton, near Montrose, Kansas. These bluestem prairies were ungrazed but were mowed in fall for the crop of hay. They were entirely representative of the area as a whole, considerable parts of which were still unbroken prairie and pasture land. They are in the Chernozem zonal or climatic soil group. To conserve space, the changes in the Carleton prairie (which is representative of the others) will be outlined.

The Carleton prairie was a nearly level upland prairie of about 80 acres; a broad, shallow ravine runs through its entire length. The soil is Crete silt loam. Pre-drought vegetation consisted mostly of little and big bluestem, with only small amounts of western wheatgrass, side-oats grama, and the short grasses—blue grama and buffalo grass. The yearly record of change follows.

1932/1933. Moderate precipitation. Little bluestem furnished 50 to 85 percent of cover; big bluestem occupied 50 percent of the lowlands but 20 percent elsewhere. Wheatgrass was nearly confined to ravines. There were only a few patches of buffalo grass and blue grama.

1934. Extreme drought. Examined in 1935.

1935. Normal precipitation. Little bluestem suffered 95 percent loss; big bluestem lost 10 to 15 percent. Less than one-fourth of the prairie was covered with a greatly thinned stand of bluestems; one-fourth was mixed bluestem and wheatgrass; and one-half was dominated by nearly pure but open stands of wheatgrass. Both short grasses had increased.

1936. Moist spring, dry summer. Some growth in spring; all grasses dried in July. Further death of bluestems.

1937. Moist spring, dry summer. Big bluestem was dominant only in local but clearly defined patches. Wheatgrass greatly extended its territory (see Figs. 85, 86). Short grasses extended their local areas and started new ones.

1938. Normal precipitation. Bluestem remained uninvaded only in small patches. Death of relict vegetation resulted in more extensive areas of wheatgrass, which had spread over nine-tenths of the prairie. Short grasses increased greatly in area.

Fig. 85.—View of the Carleton prairie in spring, 1940. Western wheatgrass (dark color) occupies three-fourths of the prairie (top), and almost complete occupation of the prairie by western wheatgrass (bottom).

Fig. 86.—Making hay in Carleton prairie in 1942. There was a fair yield but the livestock found it much less palatable than the bluestems. Consequently the sod was broken.

1939. Dry early spring, moist late spring, dry summer. Side-oats grama thrived as wheatgrass waned; former had invaded or was invading wheatgrass over almost the whole prairie. Wheatgrass ranked second. Short grasses were still increasing; pure stands were common and they were invading thin stands of wheatgrass. Also, wheatgrass was invading short grasses and thus forming mixed prairie. Area of stands of big bluestem had decreased.

1940. Dry to moist spring, dry summer. Much big bluestem had died. Former patches mixed with wheatgrass were now all wheatgrass. Only about 1 percent of side-oats grama survived. Wheatgrass increased greatly; it occupied three-fourths of prairie. Stand was thicker and there was a better cover of debris. Blue grama and buffalo grass increased greatly.

1941. Wet spring, moderately moist summer. The end of drought. Wheatgrass had invaded remaining islands of big bluestem. It thickened old stands everywhere and occupied all bared soil. It had spread widely into short grasses, especially buffalo grass; short grasses had also invaded pure stands of wheatgrass. These invasions changed true prairie to typical mixed prairie.

Almost identical changes had taken place not only at Clay Center and Montrose but generally over an area 100 to 150 miles in width on the western edge of the central portion of true prairie. The chief dominant, little bluestem, mostly or entirely succumbed to drought in 1934 and 1936. Big bluestem was greatly damaged and nearly disappeared in 1940. Western wheatgrass rapidly increased each year. In 1939 it was held in check by an enormous spread and excellent development of side-oats grama. This competitor, however, suffered great losses the following year and wheatgrass occupied most of the area. A steady increase and wide distribution of the very xeric blue grama and, in a smaller degree, of buffalo grass occurred synchronously with the rise of wheatgrass. By mutual invasions of the mid grass and short grasses, typical mixed prairie became clearly apparent in 1938. By 1941, seven years after the drought began, these grasslands were almost entirely transformed into mixed prairie (Weaver, 1943).

Where native true prairie had been weakened by grazing and trampling, or where it had degenerated to short-grass pasture, wheatgrass usually gained even earlier entrance and spread with great rapidity. This often resulted in a wheatgrass–short grass mixed prairie type.

CONTINUED DROUGHT IN THE GREAT PLAINS

The vegetation of western Kansas has been studied through seven years of continuous drought, from 1933 to 1939, inclusive. Investigations were centered at Hays, but studies were made in many counties. Water content of soil was determined weekly during the growing season and records of aerial environmental factors were obtained. Reactions of the mixed prairie vegetation have been recorded year by year in scores of permanent, widely distributed quadrats and by extensive field notes (Albertson and Weaver, 1942).

The pioneers found a dense cover of vegetation. They broke only enough sod to grow crops for their food and winter forage and grain for their livestock. Farming, which for a long time was incidental to stock-raising, received a great impetus with the invention of labor-saving machinery and a period of high prices for wheat. Despite this stimulus to agriculture, over one-third of the land remained unbroken in 1940. Saturation of the wheat market turned attention to livestock production just at a time when the range was severely depleted by drought and dust coverage. Extent of grassland was limited and overgrazing and deterioration were inevitable.

Moderately grazed and ungrazed prairies were in excellent condition in 1933 because of a very favorable six-year period just preceding, when the average annual precipitation (27.8 inches) was approximately 5 inches above normal.

Annual precipitation during each of the drought years was below normal, and during four of the seven years it was nearly 7 inches below normal. Most of this deficit occurred during the growing season. Periods of five to seven weeks with almost no rainfall occurred in summer. An accumulated deficit of 6.7 inches in 1933 increased to 21.6 inches in 1936 and to 34.5 inches in 1939.

Temperatures were abnormally high during the drought and the duration of periods with high temperatures was unusually long. Average daily maximum temperatures for June, July, and August of the four driest years were 93.2°, 100.6°, and 94.8° F., respectively—or 8.0°, 7.5°, and 4.4° higher than for the same months during 1925–1932.

Wind movement was abnormally high. That in 1934 was greatest, totaling 41,782 miles from April to September, inclusive. The lowest total, in 1936, was 33,838 miles. The highest wind movement occurred in April and May, and often resulted in great duststorms. The combined effects of cultivation, overgrazing, and drought created conditions that were extremely conducive to duststorms. Such storms were frequent but not severe during 1932. Many duststorms of wide extent occurred in April, 1933, but the blowing of dust reached a climax in March and April of 1935. Sometimes the storms were of several days' duration. Vast areas of vegetation were smothered by thin blankets of silt or by great drifts of loose earth. After the vegetation died, the dust was again moved by the wind, and thus supplied the silt for later "black blizzards" (see Fig. 87).

Fig. 87.—Approaching duststorm (in western Kansas) was typical of the "black blizzards" of the great drought. Photo by Conard.

The average seasonal evaporation (April to September, inclusive) from a free water surface was 48.5 inches preceding the drought. It was always greater during the drought and was 66.4 inches in 1934. Average monthly evaporation for the drought years was 1.8 inches higher than normal in June, 2.7 inches higher than normal in July, and 1.7 inches higher than normal in August—and 8.5 inches above normal for the growing season.

Water content of soil was determined at weekly intervals throughout the growing season to a depth of 5 feet. Available water was the limiting factor for plant production. There was about 2 percent residual water available to plants at certain depths below 2 feet in 1933, but none thereafter. Water was nonavailable in the second foot after the first week in June, 1933, nonavailable continuously (except for two weeks) in June 1934, and nonavailable during the entire month of June, 1935. From 1937 to 1939, inclusive, the second foot of soil had no water available for plant growth. Even water in the surface foot was depleted to a nonavailable level for growth during three two-week periods in 1933, one four-week period in 1934, and one seven-week interval in 1935. This also occurred when water was nonavailable at any depth to 5 feet, and the plants succumbed or became dormant. They were without available water during three separate weeks in 1937; there were similar periods of two and five weeks' duration in 1938; and in 1939 there were three such periods, including one of four weeks' duration in July and August.

Three types of vegetation, with varying degrees of intermixtures, are common in the mixed prairie of west-central Kansas. They are the little bluestem (*Andropogon scoparius*) type, common on hillsides and in shallow ravines; the short-grass (*Buchloe-Bouteloua*) type, widely distributed over nearly level uplands; and the big bluestem (*Andropogon furcatus*) type of large ravines and low, moist slopes.

During the seven years of drought, vegetation remained wilted or dried over periods of several weeks' duration (see Fig. 88). Periods of dormancy alternated with periods of growth several times during a single summer. Many mesic plants disappeared completely and even the most xeric species were reduced greatly in numbers. Animal life also was greatly depleted.

The complete story of the deterioration of the little bluestem type has been recorded. The pre-drought basal cover of about 60 percent was composed of approximately one-sixth big bluestem, and nearly all of the remainder was little bluestem. Even where it was protected from grazing, little bluestem decreased so rapidly that only 10 to 20 percent remained in 1935, and only 1 to 4 percent in 1939. Big bluestem

Fig. 88.—Blue grama on the Great Plains, at Yuma, Colo., in 1939.

was reduced to 2 percent or less. Invasion of more xeric species, especially side-oats grama (*Bouteloua curtipendula*) and blue grama (*B. gracilis*), resulted in the minimum basal cover of 16 percent in 1936 being increased to 22 to 30 percent in 1939. Accumulated benefits derived from the best drought years were often expressed by better growth the following season, even if it was dry. Myriads of seedlings sometimes appeared in the open spaces between the living bunches, but nearly all succumbed. Opportune showers permitted flowering of mature plants but few viable seeds were produced.

Late in the drought period, large areas of level prairies were found to be completely devoid of little bluestem. Some big bluestem survived because of its great depth of root system and the large amounts of reserve food in its crowns and rhizomes. Side-oats grama, once it became established in the bared areas, spread considerably, mostly by short rhizomes; the young shoots from them survived more severe drought than the seedlings. Plants of blue grama increased by tillering. The increase was slow but constant in the little bluestem habitat.

In the ungrazed open type of prairie, the pre-drought basal cover was often only 25 percent and consisted of nearly pure little bluestem. It was reduced rapidly after 1934, and to 12 percent in 1936. Increase of side-oats grama, which became dominant, increased from 4 to 11 percent, partly counterbalancing the losses suffered by little bluestem. Basal cover was about 18 percent in 1939.

Basal cover in the overgrazed little bluestem type had been reduced to 11.5 percent in 1935. Side-oats grama and blue grama constituted about half of the cover, which reached its minimum of 9.3 percent the next year. Although the bluestem all but disappeared, the total vegetation, including buffalo grass, improved steadily and attained a basal cover of 30 percent in 1939.

Studies in the Buchloe-Bouteloua type of prairie were begun in 1932 and were extended in 1935 to include several grazing treatments. In 1937, quadrats were established in pastures in ten different counties of western Kansas; some pastures had undergone various degrees of covering by dust.

Ungrazed pastures, formerly moderately grazed, and with an approximately equal mixture of buffalo grass and blue grama in almost pure stands, presented a basal cover of 80 to 90 percent. This decreased slowly until 1936, when it was only 58 percent, and the next year it decreased to 25 percent. Although fluctuating somewhat, this stand was further reduced, to 22 percent, in 1939.

Deterioration was much greater in a similar, adjacent pasture that had been ungrazed for many years. Excessive growth of the vegetation during the good years preceding the drought made it more susceptible to drought injury than pastures which had been moderately grazed. Basal cover was only 25 percent in 1935, and 11 percent the next year. Thereafter there was a slow but steady improvement, until 1938 and 1939, when it was again 26 percent. There was little change in the composition of the vegetation, although previous to 1935 buffalo grass had suffered the greater loss. In portions of the pasture there were large areas almost without vegetation. The areas that suffered the greatest losses had been composed more largely of buffalo grass than blue grama. Where local dust deposits exceeded one-half inch in depth, the short grasses were handicapped and a cover an inch or more in depth usually was fatal.

Despite great quantities of seedlings and the rapid propagation of buffalo grass by stolons in 1935 and at other times, periods favorable for growth were usually too short to result in establishment. Flower stalks were sometimes formed, but few seeds matured. With dusting and denudation, rainfall became less efficient and runoff greatly increased.

In the short-grass type that was moderately grazed during the drought, an average basal cover of about 50 percent was found in 1935.

This was reduced to about 5 percent in 1936, less than 0.5 percent being buffalo grass. It then gradually increased to 28 percent by 1939. In some areas buffalo grass, which often remained only in shallow depressions, increased more rapidly; in other areas it was the more drought-resistant codominant, blue grama. Few other species were even of minor importance.

Overgrazed short-grass ranges were reduced to a cover of 22 percent, or less than half of that, under moderate grazing. Blue grama was reduced to 2 percent, buffalo grass to less than 1 percent, and the total basal cover to about 2.5 percent when it became most depleted, in 1936. Areas 10 to 15 feet in diameter and entirely bare of grass were frequent. Recovery was aided by reduced grazing pressure, and basal cover increased year by year to 7, 13, and 19 percent.

Basal cover in ungrazed, moderately grazed, and overgrazed short-grass ranges (where the cover in 1932 varied from 88 to 80 percent, respectively), was reduced in a somewhat parallel manner. Maximum reduction was least in 1933 in ungrazed grassland. Percentages of cover were, in order, 65, 48, and 22 percent in 1935, but 58, 7, and 3 percent in 1936. By 1938 the cover was 31, 22, and 13 percent, respectively. Further heavy losses on overgrazed ranges were recorded in 1940.

The transitional area between the little bluestem and short grass exhibited several mixtures. Even where large amounts (40 percent) of little bluestem were found, total loss of the species usually occurred. Total basal cover was reduced from 65 to 25 percent. Owing to the entrance of side-oats grama and to a great increase of short grasses, about 50 percent cover was present in 1939.

Drought over most of the range area, except where it was accompanied by dust burial, resulted in severe thinning of the short grasses rather than in their complete destruction.

Side-oats grama was intermixed with short grasses in more xeric situations in the ecotone before the drought. Where it constituted 60 percent of the cover in 1932, it decreased to 3 percent by 1936, and scarcely gained thereafter. The short grasses, about equally intermixed, were more resistant to desiccation. They decreased from 22 to 11 percent, but by 1939 they had doubled their original basal area. Thus the total basal cover of 82 percent (1932) was reduced to 14 percent in 1936, but it was 50 percent in 1939.

Where several mid grasses, including wire grasses (*Aristida purpurea* and *A. longiseta*), were intermixed with short grasses in the transitional area, all disappeared in one to three years. The short grasses were at a low ebb in 1936, but subsequently they regained the ground lost by the mid grasses, so that the initial basal cover of 34 percent was again attained.

Great changes also occurred in the big bluestem type, although decrease in available soil moisture was slower here—as was also the decrease in abundance of vegetation. Western wheatgrass, side-oats grama, and other more xeric grasses partially replaced big bluestem, switchgrass (*Panicum virgatum*), and other mesic species. A loss of one-fourth of the vegetation resulted from drought and consequent soil erosion and deposit.

Intensive studies of ranges were made in ten additional counties in western Kansas, beginning in 1937. All were in the short-grass type. On lightly dusted and moderately grazed ranges, basal cover varied from 10 to 33 percent. Percentage of buffalo grass usually averaged higher than that of blue grama. Variations in cover were usually closely correlated with the amount of rainfall. No permanent gains occurred by 1939.

Under light dusting and overgrazing, a cover of 5.5 percent increased to 18 in 1938, and to 26 percent in 1939. At some stations this resulted from an excellent growth of buffalo grass. Where heavy dusting and moderate grazing occurred, and buffalo grass and blue grama were about equally represented, the basal cover of 6 percent increased to 15 percent.

Vegetation suffered the greatest losses under heavy dusting and overgrazing. An average basal cover of 1 to 2 percent showed no increase by 1939. This much-depleted type of grassland is of very wide distribution and constitutes a large portion of the vegetation near the center of the "dust bowl."

Drought, overgrazing, and hordes of grasshoppers caused great reduction in the carrying capacity of the range. Yield of palatable forage in overgrazed pastures was less than 10 percent of that produced in well-managed ones. Where 10 to 12 acres formerly were required to sustain one animal unit, 30 to 50 acres were now needed. Several years of normal or above-normal precipitation and the most judicious range management were required to restore the former cover of grasses (Albertson and Weaver, 1942).

RESPONSE OF ROOTS TO THE GREAT DROUGHT

In true prairie where the original population was little affected above ground, root habits remained normal. In prairie areas that were badly denuded by drought in 1933 and 1940, and where only widely scattered relicts or invading grasses occurred, much space between the plants was almost free from living roots. In half-bared places where relict grasses

Fig. 89.—New growths of *Kuhnia eupatorioides* (left), *Liatris punctata* (center), and *Amorpha canescens* (right) after a 10-month interval during which the upper 4 feet of the taproots had been encased in dry sand. Photo was taken Aug. 1, 1935.

or invading species had increased, depth of roots corresponded with that of current rainfall penetration, about 2 to 2.5 feet.

Very deeply rooted prairie forbs developed normally even after available moisture occurred only below 4 feet. This was ascertained by excavating and encasing the upper 3 to 5 feet of taproots in galvanized iron cylinders that had been filled with dry sand in the fall without disturbing the deeper roots (Fig. 89). Eleven plants of six species were used. Growth of the tops the following summer was equal to that of undisturbed plants (Nedrow, 1937).

In mixed prairie, bisects in the upland short-grass type revealed that roots of *Bouteloua gracilis* and *Buchloe dactyloides* were fewer in number and shallower in depth. A former root depth of 4 to 5 feet at Hays, Kansas, had been replaced by one of about 18 inches. Under lighter precipitation westward, the depth was rarely more than 12 inches. Most old roots of forbs had died. The few that grew during the drought were much shallower. In ravines in mixed prairie, which received additional water from runoff, the roots grew deeper as drought increased. Here wheatgrass, side-oats grama, and big bluestem extended their roots another foot downward and branched more profusely in

Fig. 90.—Typical view on a depleted range of the Great Plains in 1939. The clumps of grass are quite as widely spaced throughout as in the foreground. Photo was taken in eastern Wyoming.

the deeper soil than before the great drought. Many-flowered aster (*Aster ericoides*), Pitcher's sage (*Salvia pitcheri*), and ironweed (*Vernonia baldwini*) doubled their pre-drought root depth (Weaver and Albertson, 1943).

Survey of Ranges near the End of Drought

A survey was made in the summer of 1939 of 88 ranges that had been selected as representative of grazing lands in western Kansas and Nebraska, portions of southwestern South Dakota, eastern Wyoming and Colorado, and the panhandle of Oklahoma (Weaver and Albertson, 1940a) (see Figs. 90, 91).

Severe drought, overgrazing, burial by dust, and damage by grasshoppers had resulted in greatly reducing the cover of range grasses. This portion of the mixed prairie had almost completely lost its upper story of mid grasses on non-sandy lands. The short grasses and sedges had undergone a process of thinning which had resulted in only the most vigorous plants remaining alive. Many of the less xeric forbs had disappeared, and only six or eight of the most xeric native forbs were regularly represented by much-dwarfed and widely spaced individuals.

Fig. 91.—Short-grass range in eastern Colorado that is somewhat depleted by drought.

The basal cover of grasses was 21 percent or more in only 16 percent of the ranges. In another 16 percent it ranged between 11 and 20 percent. It varied between 6 and 10 percent in 28 percent, and was reduced to 2 to 5 percent in another group that totalled 16 percent. The remaining one-fourth of the pastures (24 percent) presented a cover of 1 percent or less.

Extremely poor conditions varied with the better ones throughout. The bare soil during periods with moisture was populated with annual weeds, chief of which was Russian thistle. In many places it was only with difficulty that one could distinguish denuded pastures from weedy, tilled land. Cacti had increased greatly almost everywhere and constituted a serious problem. Because of the low precipitation of 1939, most ranges had lost any gains that had been made during favorable periods since 1934, and further reduction in vegetation seemed certain if the winter also was dry.

In 1956, during the recurrence of another severe drought, Tomanek and Albertson (1957) found that buffalo grass was the most abundant increaser of all the grasses in western Kansas. In the northwest it increased from 16 percent on ungrazed ranges to 90 percent, and in the southwest from 5 to 90 percent. This low-growing species increases tremendously on all sites, under both moderate grazing and heavy use. Its tendency to increase rapidly with grazing is made possible by the fact that the growing point is in the ground (where it occurs in the surface soil) and is not easily injured (Branson, 1953). Also, the rapid growth of its stolons permits vegetative spreading. Under favorable conditions, stolons may elongate at the rate of an inch or more per day, and an average spread of 8 inches or more may occur during a single season.

The rapid growth of buffalo grass has made it valuable in quickly providing a cover on an area denuded by drought, dusting, or overuse. Although a large percentage of this grass generally indicates over-utilization, it is still very valuable on the Great Plains. It is relished by cattle and it preserves more of its nutrients in winter than most of the taller grasses. Thousands of acres which were formerly covered with a mixture of mid and short grasses are now almost pure buffalo grass ranges. Such ranges are fine for the protection of the soil, and they furnish excellent forage for livestock, but when they reach this stage of degeneration production is low compared with a mixture of short and mid grasses. Experiment reveals that, under moderate grazing, twice as much forage is common (Tomanek and Albertson, 1957).

Moreover, the excessively grazed buffalo grass has lost most of its vigor. It has been ascertained that a moderately grazed pasture may produce 3 times as much forage as such a heavily grazed one (Tomanek, 1948). Tomanek and Albertson (1957) pointed out that in terms of practical range management, the rancher could allow the livestock to remove half of the forage produced on the moderately grazed land, and leave half to protect the soil, to conserve soil moisture, and to store food in the roots for further use, and still get as much forage as if the livestock ate all of the grass on the heavily grazed range.

The effects of the very damaging drought cycle of 1952 to 1955 on the Great Plains was studied by Albertson (1957) and others. Among 24 representative ranges in various parts of six states, losses of vegetation ranged from 10 to 99 percent. The heavier losses were nearly all in ranges with a high degree of utilization; in fact, they were nearly double that of ranges moderately grazed. Grasslands weakened by overgrazing during wet cycles are extremely sensitive to deficient soil moisture when drought strikes. The researchers pointed out that the pioneers observed a higher carrying capacity of range land during wet periods, and great decreases in drought, but the contribution overgrazing made to drought losses was largely overlooked. Presumably, native vegetation developed under conditions similar to those of today, and it is also safe to assume that native plants will continue to dominate the prairies if they are not continually overgrazed by livestock or buried too deeply by soil blown from cultivated fields. Therefore, if our native vegetation is completely destroyed, man should be held accountable (Albertson, Tomanek, and Riegel, 1957).

X.

Recovery of Vegetation

For eight years, 1933 to 1940, drought had prevailed in the mid-continental grasslands. The true prairie west of Minnesota, Iowa, and Missouri had been woefully depleted. Mixed prairie farther westward, as far as the Rocky Mountains, had been nearly or entirely destroyed. But the parched soil became wet once more, wet to a depth of several feet. The hot, dry air became warm and moist. The terrible duststorms had ceased. A changed environment had come at last, ending the long period of drought. Thus in 1941 the long-delayed recovery of vegetation began. The nature of this complex phenomenon was now to be revealed and the sequence of recovery to be recorded (Albertson and Weaver, 1944).

DEVELOPMENT OF GRASSLAND TYPES

With the recurrence of years of normal or above-normal precipitation, vegetation recovered. At first the processes were slow, since deep dormancy often follows great drought, but later they became greatly accelerated. When the weedy native forbs had been subdued by the increasing population of grasses, when the annual grasses and weeds had been mostly replaced by perennial grasses, and when the dominant species through competition had replaced the former interstitial ones, there crystallized out of the heterogeneous drought populations several very definite grassland communities or types. In fact, the western wheatgrass type had grown steadily year by year almost from the beginning of the drought. Likewise, the short grasses had migrated into new territory and gradually claimed it for themselves (see Fig. 92).

Other types were less clearly defined since dominants were much intermingled; and subdominant species such as side-oats grama and Junegrass at times played a leading role in repopulating the bared soil. Even in 1940, much soil was still open to invaders in some prairies, and many places were populated only thinly by individuals of a potential grassland type. But with better conditions for growth, all bared places were reclaimed. Competition sorted out the permanent dominants from the seral or temporary ones, and the boundaries of the

177

Fɪɢ. 92.—Short grasses invading the true prairie during the drought.

communities in the mosaic of prairie patterns became quite definite. For comparison, the pre-drought upland communities and newly developed ones are listed.

Recovery of vegetation was delayed again and again throughout the seven years because losses in the driest years offset gains made during the less dry ones. Recovery from dormancy sometimes did not

TABLE 10

Cᴏᴍᴍᴜɴɪᴛɪᴇs ᴏғ Uᴘʟᴀɴᴅ Tʀᴜᴇ Pʀᴀɪʀɪᴇ (Wᴇᴀᴠᴇʀ, 1950)

Before the Great Drought *(1933)*	*Old or Modified Types* *after the Great Drought* *(1943)*
Little bluestem . .	1. Relict big bluestem–little bluestem (4)*
	2. Big bluestem . . . . (6)
	3. Mixed grasses . . . . (7)
Needlegrass . . .	4. Needlegrass . . . . (3)
Prairie dropseed . .	5. Prairie dropseed . . . (8)
	Newly Developed Types
	6. Western wheatgrass . . . (2)
	7. Blue grama . . . . (5)
	8. Mixed prairie . . . . (1)

*Numbers in parentheses indicate the sequence, in order of area covered, of the several communities; mixed prairie was the largest type. This order is followed in the discussion.

occur until one or more years after good rains came. Vegetation was usually slow and conservative in its changes. The degree of recovery varied with the amount and seasonal distribution of soil moisture. From 1941 to 1943, however, gains were cumulative and the prairie cover again became almost complete, even though it differed greatly from the former one.

1. A mixed prairie type was formed by the replacement of bluestems by western wheatgrass and short grasses, and by the later intermingling of mid grass and short grass (see Fig. 93). This began in 1938 and was completed in 1941. Invasion had ceased, since the remaining area was occupied by other vegetation, but not the intermixing of short grasses and wheatgrass.

2. Western wheatgrass spread by seed and appeared over a wide territory immediately following 1934; the invasion was well advanced by 1938. It was favored by moist springs and dry summers, but often it utilized all available moisture and thus caused the dwarfing or death of most other vegetation. Relict bluestems were usually replaced by pure stands of wheatgrass.

3. Needlegrass had spread so widely that it ranked third in the areas it occupied. A relict big bluestem–little bluestem type resulted

Fig. 93.—Invasion of blue grama (white patches) into bluestem prairie at Hebron, Neb., during the early part of the great drought.

from modification of the pre-drought little bluestem consociation, but it covered relatively small areas. Usually, much little bluestem had died and its place was taken by big bluestem. Some survived as dormant rootcrowns.

4. The big bluestem type on uplands resulted from the survival of this grass wherever the former little bluestem type had lost its chief component, *Andropogon scoparius*. After heavy initial losses, big bluestem survived the drought in the bunch habit, but later, unless overwhelmed by wheatgrass, it spread to form a sod.

5. Blue grama often survived where all other grasses died. It spread almost without interruption from 1935 to 1942, and promptly thickened its stands. Buffalo grass, less abundant and more greatly harmed by drought than blue grama, also increased greatly (Fig. 93).

6. Stands of big bluestem on lowlands often were greatly thinned, but usually it replaced its losses and reestablished the tall-grass type.

7. Mixed grasses formed a community of variable composition over considerable areas where widely scattered plants of different species recovered, reseeded, or invaded at about the same time. Also, big bluestem spread widely or renewed its growth from dormant rhizomes and formed mixtures with various other grasses. These species have merely completed the occupation of bared soil where no other types had gained control. The community is not climax vegetation.

8. Prairie dropseed, of minor importance in pre-drought years, re-covered from extensive early losses and extended the area of its com-munities greatly. Side-oats grama, Junegrass, and several other grasses which were pre-drought interstitials became important seral dominants. Some were temporarily so abundant as to threaten the dominance of wheatgrass and big bluestem, but all have been reduced in amount.

METHODS OF RECOVERY

The methods for recovery of the grasses were increase in size of relicts by tillering, production of rhizomes and stolons, awakening of old crowns and underground parts from deep dormancy, and produc-tion of an abundance of seedlings.

The ability to adjust itself to the environment by various degrees of tillering accounts in a large measure for the successful occupation of more of the earth's surface by grasses than by any other lifeform. The very large size attained by relict bunches of various grasses after the early drought illustrated their adjustment to an increased water supply resulting from decreased competition. Thus plants of side-oats grama, prairie dropseed, and other bunch grasses developed to great size despite the usual scarcity of soil moisture.

The extremely rapid spread of big bluestem in 1938 not only into open ground but also among bunches of prairie dropseed, little bluestem, and elsewhere was due to rhizome propagation. The great abundance of this species, second only to western wheatgrass, is due first to drought survival because of its very deep root system and food stored in rhizomes, but thereafter to its rapid spreading everywhere by rhizomes. Wheatgrass, among all the grasses concerned, is more efficient in this method of invasion than bluestem. This is due in part to its resumption of growth at lower temperatures in spring and consequent evasion of drought.

The breaking of a long period of dormancy in crowns and rhizomes of grasses was observed in some prairies after a single year of good rainfall, and in nearly all prairies, including areas of western wheatgrass, in 1942 and 1943.

Little bluestem made some recovery in the spring of 1939. In general, it began to break its long dormancy and to rejuvenate from old crowns in 1942, but only after a year of good rainfall. In 1943 it revived in certain places on dry south slopes where it had not been seen since 1934. At first, only a few stems appeared, even where large bunches had grown before. Often, the remainder of the crown was dead. Such

Fig. 94.—Result of a prairie fire near Hebron, Neb., in July, 1939. Note the low stature of the prairie grasses in this dry year.

tufts could easily have been mistaken for new bunches developed from 1941 seedlings; actually, however, there were but few of these (Fig. 94).

Big bluestem also survived long periods of dormancy. In one Kansas prairie in 1942, for example, a fine open stand of big bluestem appeared from old crowns and rhizomes which had been dormant for several years. It occurred mostly in the interspaces between the bunches and sods of blue grama, but its vertical stems sometimes penetrated through them as well. In the Pleasant Dale prairie near Lincoln, 20 sampling plots were selected in 1943 in the wheatgrass community, which had been well established since 1935. Care was taken to select them where the wheatgrass was pure, since they were to be used to determine production of this grass compared with that of big bluestem. No big bluestem was found in any of them at this time, or when the grass was clipped on May 19 and again on June 28. Nor was big bluestem observed in closely adjacent prairies where other plots were selected for later clipping when the grasses had matured.

On August 16 to 20, when a third cutting was made, considerable big bluestem had appeared in several of these plots. It was 5 to 15 inches tall, scattered lightly to rather thickly, and it occurred widely in places where in May and June no bluestems had grown. This is exact evidence of the reappearance of a grass which had lain dormant either since 1934 (9 years) or certainly since 1938 (5 years) without appearing above ground. In most plots, only a single stem or a few stems were found in a place, but in some it appeared as if the remnants of widely spread sod were growing. Consequently, several large blocks of sod were removed to a depth of 4 inches and brought to the greenhouse for study. When the tops were cut an inch above the soil and the soaked soil was slowly and carefully washed away, the skeleton of extensive crowns and rhizomes was found.

The roots had nearly all decayed, but the rhizomes were distinct, not as continuous stems but as dead, corklike pieces 0.25 to 1 inch long. The rhizomes still were in place throughout the 4- to 12-inch interspaces between the living stems. The new shoots came not from seeds but from the largest and best-preserved rhizomes. The roots and tops were new but the basal portions were distinctly old rhizomes that had decayed away from the growing point. Numerous rhizomes were found which were undecayed, but apparently they had no vitality. A few were seen which had a single viable bud that had just begun swelling to begin growth. Many of the survivors were clearly from previous bunches and only a few inches in diameter. Big bluestem revived from dormancy even where wheatgrass, prairie dropseed, or blue grama had invaded a tract and grown for several years. It was also found that

FIG. 95.—Recovery of little bluestem from old crowns that retained some life during the 7 years of the great drought.

Sorghastrum nutans, Sporobolus heterolepis, and certain other species re-mained dormant for a period of years (see Fig. 95).

Seeds of western wheatgrass, needlegrass, blue grama, and a few other grasses produced seedlings that became established during years of less-severe drought. Little bluestem and big bluestem produced no seed or seedlings, or seeded only rarely, until 1941. Seedlings were abundant in bare places in 1942 and 1943, and plants only 2 to 3 years old were common in 1943.

Recovery was expressed by an increase in numbers, vigor, and stature of plants, and by the resultant thickening of the stand and the exclusion of ruderals. It was also shown by the reappearance of societies of forbs, reconstruction of a layered vegetation, accumulation of debris, and increased yield.

Forb populations were not replaced during drought. Seedlings were rare until 1941 and 1942. Many species grew—sometimes thickly, after a year or two of good precipitation—from seeds that had been dormant for 7 years. Others came from long-dormant crowns, rhizomes, or roots.

Vegetation developed remarkably after the drought. In ravines where big bluestem, Indian grass (*Sorghastrum nutans*), and switchgrass

Panicum virgatum) apparently had been replaced by wheatgrass, they not infrequently reappeared from rhizomes. In 1943 in an experimental area, big bluestem was suppressed by a thick growth of prairie cordgrass (*Spartina pectinata*) that grew from dormant rhizomes. Yields on uplands increased from 0.79 ton per acre in 1940 to 1.85 tons in 1941. Yield in 1943 was 2.15 tons per acre.

RECOVERY IN MIXED PRAIRIE

Recovery in mixed prairie, where most of the area is range land, was delayed by overgrazing in some places and by understocking in others, but usually it was prompt. Major features of recovery were closely connected with the development of buffalo grass, blue grama, and sand dropseed (*Sporobolus cryptandrus*). Buffalo grass entirely disappeared from some ranges, and in many it remained alive only in the best-watered places; but other ranges retained small tufts or occasional patches. Blue grama was never killed as uniformly nor as completely over a wide area. Unless deeply buried by dust, some bunches nearly always remained. Sand dropseed was widely spread but was sparse before the drought; it increased greatly during the late phase of the drought and thereafter.

Rates of recovery varied widely, depending upon the kind of pre-drought cover, degree of depletion, and kind of grass relicts at the end of the drought. Other factors were the amount of damage from burial by dust, intensity of grazing and trampling during recovery, and amount and distribution of local precipitation. Recovery was always slower where blue grama alone remained—slower than where buffalo grass was intermixed with it or was the sole survivor.

Study of percentages of composition showed that the pre-drought sod of about equal amounts of buffalo grass and blue grama was replaced after the drought by sod with two-thirds buffalo grass and one-third blue grama. But in thinner, drier soil of an equal pre-drought mixture, the more xeric blue grama finally composed about three-fourths of the cover. Where little bluestem formed 45 percent of the cover and side-oats grama 17 percent before the drought, the post-drought cover contained 3 percent little bluestem, 66 percent side-oats grama, and 21 percent short grasses. Big bluestem nearly regained its 76 percent of cover in its type, after being reduced to 15 percent, with proportional decreases in invading wheatgrass and short grasses.

Succession on badly or completely denuded ranges consisted of four stages. The first weed stage was composed chiefly of a dozen annual forbs, among which were Russian thistle, lamb's quarters, narrow-leaved goosefoot (*Chenopodium leptophyllum*), common sunflower (*Heli-*

anthus annuus), monolepis (*Monolepis nuttalliana*), and horseweed. They occurred wherever the soil had been laid bare by drought or wind erosion or deposit. Complete destruction was widespread, especially in the southwest. They played an important role in temporarily stabilizing the soil.

A second weed stage consisted of little barley, six-weeks fescue, peppergrass, species of plantain and stickseed (Cryptantha), and other species which usually dominated more stable soil than the preceding one. These early-maturing plants formed either a discontinuous patchy layer above the grasses or overshadowed them entirely throughout the range.

An early native-grass stage consisted of sand dropseed, western wheatgrass, tumblegrass (*Schedonnardus paniculatus*), false buffalo grass (*Munroa squarrosa*), and windmill grass (*Chloris verticillata*). By 1940, sand dropseed was the most widely spread and the most abundant. Thickness of stand was somewhat proportional to the degree of destruction of the short grasses. The sand dropseed type became common. Wheatgrass was represented widely but often sparingly, except in depressions or on rolling land.

The late grass stage, consisting chiefly of short grasses with wire grasses (Aristida), side-oats grama, and squirrel tail (*Sitanion hystrix*), was the same in both subseres.

Development of the late grass stage was marked primarily by the return or increase of blue grama and buffalo grass. These grasses became—as before the drought—the most abundant grasses of the midwestern ranges. Mid grasses, scattered throughout before the drought, nearly disappeared. With rare exceptions, they were only beginning to reappear.

Relict blue grama, often covering only 1 to 5 percent of the soil, increased rapidly after 1940. Development of both seedlings and tillers was often retarded by too great a reduction of light, which was caused by dense stands of weeds. Where the drought was most severe, nearly pure stands of blue grama often developed.

Buffalo grass, somewhat less drought-resistant than blue grama, was more easily smothered by dust. Where tufts or seedlings were widely distributed, recovery was much more rapid than that of blue grama, and this resulted in a great preponderance of buffalo grass in the new cover. Intermingling of the two short grasses and the formation of dense, uniform cover resulted.

Cover in post-climax vegetation that had been dominated by big bluestem was greatly thinned by drought and dormancy of the tall grasses. The area was invaded by coarse weeds; certain mid grasses

Fig. 96.—A thick stand of cactus in northwestern Nebraska.

increased greatly; and even short grasses invaded the area. After the drought, these were largely replaced by big bluestem. Little bluestem, which dominated on steep hillsides, was reduced from about 45 to 1 percent; its place during the drought was taken by side-oats grama and blue grama. It is returning slowly.

Recovery from drought was clearly expressed in increased yield. Production in the big bluestem type in 1942 was 2.4 times as great as in 1940; in the moderately grazed short-grass type, production increased 3 times; and in the overgrazed short-grass type, where buffalo grass was abundant, the increase was 19 times as great.

All but 12 species of forbs disappeared from the short-grass ranges and all but one of these were scarce. Increase after the drought was from dormant underground parts, from dormant pre-drought seed, and from seed from relict plants. Some species showed no recovery; only a few occurred in normal abundance.

Cactus increased so greatly during the drought that it hindered utilization of the grasses (see Fig. 96). With increased soil moisture, a thickened cover of grasses, and inceased humidity about the cactus, its insect enemies thrived and the cactus was more or less reduced to the amount in which it occurred before the drought (Weaver and Albertson, 1943).

DROUGHT RESISTANCE OF SEEDLINGS

Studies of the severe damage to native midwestern grasses as a result of the drought was fundamentally an investigation of the drought resistance of the various species of dominant grasses and accompanying forbs. Since the most critical stage in the life of a plant is usually that of the seedling, final experimental evidence of relative drought resistance was sought by growing large numbers of seedlings in the greenhouse under the same conditions of soil drought or of soil and atmospheric drought.

Seeds were sown in hills an inch apart. Several seeds were placed in a hill and plants were thinned to 1 per hill shortly after emergence. Species were arranged so that, somewhere, a plant of a particular species was adjacent to plants of each of the other species.

Mixed plantings of seedlings of 14 species of dominant prairie grasses were tested for resistance to drought at ordinary summer temperatures. Only a few plants, all of which were short grasses, survived where the drought was most critical. Of these, blue grama showed the greatest drought resistance; 3 times as many blue grama survived, compared to hairy grama (*Bouteloua hirsuta*) and buffalo grass, which were next in order. Western wheatgrass was least able to resist drought.

There was no mortality and only slight damage to seedlings that were subjected to hot winds of 135° F. when soil moisture was available. Leaves of the short-grass seedlings (listed above) were scarcely injured by temperatures as high as 145° F. Differences of a few weeks in age and of previous exposure to drought had no significant effect on the survival of seedlings which were exposed to hot winds.

When both the results of exposure to soil drought and to atmospheric drought were taken into account, *Bouteloua gracilis* was the most drought resistant. In order of decreasing drought resistance, the species were *Buchloe dactyloides*, *Bouteloua hirsuta*, *Sporobolus asper*, *Bouteloua curtipendula*, *Stipa comata*, *Andropogon scoparius*, *A. furcatus*, *Panicum virgatum*, *Sorghastrum nutans*, *Stipa spartea*, *Koeleria cristata*, *Elymus canadensis*, and *Agropyron smithii*. In general, species that are characteristic of uplands or that normally occur westward in mixed prairie were more resistant; species found in the lowlands or in the true prairie were less resistant to drought (Mueller and Weaver, 1942).

ROLE OF SEEDLINGS IN RECOVERY

An extensive survey of the degree of deterioration of midwestern ranges was made in 1939 (Weaver and Albertson, 1940a). Decreases in the basal cover of the grasses from extreme drought, overstocking, and damage by dust were appalling. Basal cover by the native grass had decreased to 1 percent in one-fourth of the ranges examined, and it was 21 percent or more in only 16 percent of the ranges. Little, if any, improvement occurred during the following year; instead, certain large areas suffered distinct losses as a result of a severe late-summer drought. The perennial grasses over large areas of southwestern Kansas, for example, had been reduced to a basal cover of about 3 percent in 1937. The cover improved slightly during the two following years, but it was reduced to less than 0.5 percent during the very dry fall of 1939 (Albertson, 1941). On some of the better ranges, where the basal cover was about 15 percent, scarcely a spear of grass remained (Fig. 97).

The drought continued, until 1941, with only slight interruptions for eight years. Production of viable seed by the native grass had usually been prevented by dry weather that rapidly followed a period of flowering. Consequently, it has seemed that the seed supply in the soil might have become greatly depleted and that a single growing season would not be sufficient for greatly improving the range as regards the growth of new plants. Exceptions were buffalo grass, western wheatgrass, and a few other species which have excellent means of propagation by stolons or rhizomes. Consequently, it was planned—early in the spring of 1941—to make a survey of the viable seed population in the surface soil of these depleted grasslands. Later, a quantitative study of the seedling population also was made. Fortunately, 1941 was a year of abundant rainfall and consequently was very favorable to seed germination.

Seeds of range grasses, with rare exceptions, are found on or near the surface of the soil; therefore, samples of soil were taken only to a depth of 0.5 inch. Each sample included the topsoil in an area of 1 square foot. With few exceptions, five 1-square-foot samples were obtained from each pasture or range. The wide range of sampling included 7 counties in Nebraska, 4 in eastern Colorado, and 17 in Kansas. Altogether, 196 samples were obtained late in March or early in April, at a time when no seedlings were found.

An environment favorable to the germination of seeds was maintained in the greenhouse while the study was in progress. Seedlings were counted, recorded, and removed from their plots as quickly as they could be indentified. Although distinguishing characteristics appeared relatively early in the grasses, verifications were not made before 3 or more leaves had developed.

Seedlings grew at the average rate of 67 per sample. Forty species of forbs occurred, of which more than 96 percent were annual weeds, mostly *Amaranthus* spp. and *Salsola pestifer*. There were 26 species of grass

Fig. 97.—Bared ranges where samples of soil for seeds were taken: Syracuse, Kan. (top); Dighton, Kan. (center); and Cheyenne Wells, Colo. (bottom).

seedlings, of which 20 percent were ruderals, mostly *Eragrostis cilianensis*, *Hordeum pusillum*, and *Panicum capillare*.

Seedling grasses (6,730) slightly exceeded the number of forbs (6,388). More species of grasses grew in samples from true prairie than from mixed prairie, but the reverse was true for weedy forbs. Seedlings of native perennial forbs were very few.

Three species of native perennial grasses, *Sporobolus cryptandrus*, *Bouteloua gracilis*, and *Buchloe dactyloides*—all of high forage value— furnished 73 percent of all the grass seedlings. Seedlings of other perennial grasses were largely confined to true prairie stations and were relatively unimportant.

The ranges from which the soil samples for seed germination had been secured were studied between June 10 and June 26. The soil was examined in regard to water content and consequent suitability for seed germination; the composition and the development of the vegetation were studied; and the abundance of the seedlings was ascertained.

The spring was cool, late, and—like early summer—unusually wet. The total precipitation not only was well above the normal at every station but often it was several inches above—and not infrequently 2 to 4 times the normal amount. Moreover, the showers were so well distributed that soil moisture was unusually favorable for the germination of seeds and the establishment of seedlings. Numerous days with only a trace of rain or light precipitation indicated much cloudy weather. No prolonged periods of drought occurred. These data showed conclusively that, on almost every range, a continuous supply of moisture had been available to promote germination of seeds and establishment of seedlings.

Numbers and kinds of seedling grasses were determined in June in each of twenty-five 1-square-foot areas in each of 22 ranges or prairie from which samples had been taken. Extensive soil sampling and study of numerous rainfall records showed that an almost continuous supply of moisture had been available to promote germination and establishment of seedlings.

Twenty-four species of seedling grasses were found in true prairie, but only 7 in mixed prairie. Mixed prairie included about two-thirds of the places examined, but only 29 percent of the total seedlings were found here.

Of the 550 square feet of soil on which seedling grasses were counted 37 percent of the area did not support any seedling grasses. Seedling grasses were especially rare in drought-stricken and dust-covered ranges of western Kansas. Seedlings of *Sporobolus cryptandrus* and *Bouteloua gracilis* were the most widely distributed and most abundant of the perennial grasses. Viable seeds of native perennial forage grasses, with rare exceptions, were present in such small numbers (26 per square foot) as to be of very limited value—when seedling hazards are considered—in the restoration of the vegetation.

Average distribution of perennial grass seedlings on the ranges and true prairie was 4.3 per square foot; in mixed prairie alone, the average distribution was 2.4 per unit of area. Even if all of the seedlings (exclusive of the stoloniferous buffalo grass) had survived and had made a maximum growth in the mixed prairie, they would have increased the cover by less than 2 percent.

Several years of good seed production and of development of seedlings into mature grasses—along with very judicious range management—are necessary for the restoration of midwestern ranges (Weaver and Mueller, 1942).

FINAL RECOVERY AND STABILIZATION

Between 1941 and 1943, after the seven years of devastating drought and after the grasslands had suffered a terrible catastrophe, the prairie had succeeded in repopulating most of the bared soil. This was accomplished largely by a tremendous increase of its most drought-evading and drought-resisting perennial species. However, a series of changes in the vegetation seemed inevitable before the pre-drought stability could be attained.

These changes included an increase in the former dominants, the suppression of drought dominants under the wetter phase of the climatic cycle, the disappearance or suppression to normal numbers of other relict drought populations, and an increase of drought-depleted forbs. Other necessary changes were a new development of societies, reduction in height-growth with the return of increased competition, reestablishment of the understory, and the building of a soil mulch. These phenomena were studied during a period of five years that were very favorable for stabilization (Weaver, 1950*a*).

After 1943, needlegrass spread so widely that it formed the third most extensive community. Prairie dropseed also increased its territory enormously, but it still ranked lowest in extent among the eight prairie communities.

Little bluestem, formerly the most abundant of all the prairie grasses, formed the most extensive grass of upland communities. But upland losses were so severe and readjustments so great that, by 1943, the largest portion was occupied by a relict big bluestem–little bluestem type; the remainder was occupied by a big bluestem type and by mixed grasses.

In addition to these five old or modified types, three new types— western wheatgrass, blue grama, and mixed prairie—had developed. The area occupied by mixed prairie was largest, the area occupied by western wheatgrass was second-largest, and the area occupied by needle-grass was third-largest in the greatly disturbed true prairie west of the Missouri River.

The persistence of communities of wheatgrass and their marked effect in modifying the water relations of the soil were most impressive phenomena. Wheatgrass had not extended its area since the bared soil was reclaimed by other perennial dominants. Only rarely had it surrendered its control over very extensive areas it had occupied during the drought. Wheatgrass, unlike other prairie grasses, develops a very inefficient mulch. Often the soil remained nearly bare. Hairy chess (*Bromus commutatus*), a chief species of the understory, often formed a good mulch. An understory of blue grama also often occurred. A few prairie grasses invaded wheatgrass areas in wet years, but competition for water was too severe to permit a general invasion.

The presence of wheatgrass retarded rainfall penetration or the infiltration of water applied artificially, often to the extent of 35 to 50 percent. Extensive sampling had shown that the soil is always much drier under a wheatgrass sod than under adjacent prairie grasses. The bare soil or meager mulch permits beating rains to destroy the soft, crumblike soil aggregates at the surface. Fine particles are carried into the soil with the water and more or less completely clog the pores in the surface layer.

Competition of wheatgrass was so severe in the drought that other grasses, and even very deeply rooted forbs, were dwarfed or killed. Dwarfing of forbs still persisted. The difficulty in the establishment of seedlings was due to the large amounts of water that were absorbed and transpired by this early, deeply rooted, sod-forming species, which is also active early in spring and late in autumn.

Where very open stands of big bluestem alone remained on uplands after little bluestem had died, invasion by wheatgrass usually resulted in suppression of the bluestem. However, where big bluestem was more abundant and firmly entrenched but dormant, invasion by wheat-grass was sometimes unsuccessful. Any progress in the replacement of wheatgrass in upland areas had been slight.

The needlegrass type was unlike that of the pre-drought community in the absence or small numbers of intermixed prairie grasses, a paucity of forbs, the absence of an understory, and the lack of a continuous mulch.

An indicator of delay in recovery was the persistence in places of certain species in the marked overabundance they had attained when the soil was otherwise bare. Chief among these were *Aster ericoides*,

Erigeron ramosus, Solidago glaberrima, Oxalis violacea, and *Anemone caroliniana.* They are examples of similar species with organs for the underground storage of food. Examples among the grasses were side-oats grama and purple lovegrass (*Eragrostis spectabilis*), and the lingering presence of sand dropseed (*Sporobolus cryptandrus*) and hairy chess.

The great upsurge of vegetative development of practically all prairie plants during and following the very wet year of 1944 was pronounced. The spreading of big bluestem was marked in portions of all of the prairies, except where western wheatgrass thrived. Parts of dormant crowns resumed growth as single- or few-stalk plants. The stand at first was open, but widespread propagation by rhizomes was rapid and the sod gradually thickened. The greater height, deeper roots, and sod-forming habit of big bluestem placed the bunch-forming prairie dropseed and the low-growing blue grama at a great disadvantage; short grass was completely ousted in three years.

Little bluestem, after the drought, was of only intermediate importance and was outranked in abundance by six other dominant species. But flower stalks developed, seeds ripened, and seedlings and small tufts were common by 1943. Its return in abundance—often from crowns which had been dormant for five to seven years—in many drought-populated places and in relict areas was spectacular. Differences in relict stands and newly populated areas have been studied.

Bluegrass (*Poa pratensis*) was almost absent by the end of the drought though formerly it had composed about 5 percent of the vegetation. Recovering slowly in ravines, it spread upward along them onto the hills. In wet years it completely covered much bare soil or formed a thin understory to whatever prairie grasses were in possession (except wheatgrass, blue grama, and prairie dropseed). In 1950 it formed many large patches, and often exceeded its pre-drought abundance. It is important in restoring the soil mulch.

Conditions for the growth of seedlings after the drought were excellent. In 1943, those of most grasses developed into small bunches. Next year they produced abundant seed and new vegetation flourished. Seedlings of forbs were rare immediately following the drought. By 1944, those of certain species were plentiful in places. Reseeding of interspaces among the thickening vegetation continued throughout the succeeding years.

A good soil mulch occurred everywhere in the pre-drought prairie, but almost none was found on the bare, black soil during the seven years of drought. By 1944 the denser vegetation had gone far toward reestablishing a protecting cover of debris. Although every species contributed to this process, most efficient were Junegrass (*Koeleria cristata*), prairie dropseed, bluegrass, blue grama, side-oats grama, and hairy chess. The mulch was often only partial under needlegrass and western wheatgrass and in open stands of the mixed-grass type.

Forbs increased greatly in size three years after the drought, but these were nearly all old plants of the relict bluestem type. Here, too, they first increased in numbers. Elsewhere, societies of forbs were few and highly localized in the more favorable sites. Many species were not found until 1944, and some returned even later. By 1945 it was clear that certain species had spread widely and grew abundantly in sites where they did not occur before the grasses had been greatly depleted. In general, the forb population thickened to form societies only slowly. Even seven years after the drought, the layer of understory vegetation was still far from complete.

The return of the prairie to its pre-drought condition was traced year by year until the fall of 1960 (Weaver, 1961). Two decades have been required for the replacement of certain drought dominants, such as *Agropyron smithii* and *Bouteloua gracilis*, and for the reduction of *B. curtipendula*, *Koeleria cristata*, and others to their usual place as interstitial species. *Panicum scribnerianum*, *Carex pennsylvanica*, and similar species have returned to their former place in the understory. The return of the dominant grasses—little bluestem, big bluestem, needlegrass, and prairie dropseed—and the expanding or shrinking of their former holdings have been considered. The disappearance of the relict drought populations and the decrease of *Aster ericoides*, *Solidago missouriensis*, and others have been described. The long-delayed and slow restoration of the many forbs and the later development of societies have been discussed. The thickening of the plant cover and the regulation by competition of its height-growth occurred slowly through the years. Finally, the reestablishment of an understory, the seasonal aspects, and the production and maintenance of a good vegetal mulch have been attained. Thus, gradually over a period of 20 years (1940 to 1960) and under the control of soil and climate, the same kind of prairie vegetation that existed before the seven years of devasting drought has slowly been replaced.

The great drought has fully demonstrated the fact that development is the key to an understanding of vegetation. As was pointed out by Weaver and Albertson (1956), a study of a community at a particular time and place is in itself not sufficient for a complete comprehension of its significance. The continuous tracing of succession and other changes over long periods of time has aided greatly in understanding the dynamic flow of processes and the fact that vegetation is not static but always undergoes change.

XI.

Origin, Composition, and Degeneration of
Native Midwestern Pastures

When the settlers came to the Midwest, they entered an almost boundless area of grassland. Throughout the years most of the prairie has been broken for crop production, but many tracts of native vegetation remained in the western portion of that part of the vast grassland that is composed of tall and mid grasses and commonly known as true prairie. Most of it had been broken for crop production by 1920. Some areas were mowed annually for their crops of excellent winter forage, but far greater ones were used as native pasture. One result of the long years of study of the grazed portion of this grassland was the recognition of the manner in which prairie degenerates under continued grazing. A second result was the classification of the resulting pastures into several grades or range conditions, each based upon the composition of the vegetation.

When the prairie is grazed and trampled, various changes occur, and the nature and extent of these changes vary somewhat with the degree of disturbance. Once degeneration is well under way, however, it proceeds so gradually and effectively that it is usually not observed until a great loss in productivity is sustained; and several years are required for recovery. The stages of degeneration, how and why they occur, the types of pasture that are produced, and the causes of complete disintegration will now be outlined and illustrated.

Climax grassland, when it is grazed lightly or moderately, may essentially retain its natural composition over extremely long periods. It is only when grazing animals are circumscribed in their range by fences, and when too large a population is thus confined, that grazing and trampling become so excessive that the normal cover cannot be maintained.

The selection of grass types and the preference for certain species by livestock is marked when forage is plentiful. Repeated partial removal of the most palatable grasses results in the better growth of the remaining vegetation. In fact, if certain favored plants are grazed too early, too often, and too closely, they will disappear entirely. Less desirable species then receive more light and increased water, as well as additional nutrients, which normally are used by the more palatable grasses. In consequence, the undesirable species flourish and may actually increase, often with marked rapidity. Thus, during the early stages of grassland degeneration there is considerable shifting of the plant population, but this happens entirely among the species which are normal components of native prairie.

As the hold of native species is weakened by continued pasturing or intense overgrazing, invaders move in. The great stability of natural grasslands and the absence of weeds have been emphasized by Weaver and Flory (1934). Under pasturing, however, many small, bare places appear. The bare areas invite invaders which, once established, furnish seed for a new population. Gradually the native grasses and forbs are partially or entirely replaced by invading species which are better adapted to close grazing and trampling. Nearly all of these are less productive, or less palatable—or both—than the original occupants. With the disappearance of most of the native population, the prairies are far advanced on their way toward final disintegration (see Fig. 98).

Fig. 98.—An old bluestem pasture in southeastern Nebraska that was badly infested by various native and introduced weeds.

CLASSIFICATION OF GRASSES, BASED UPON THEIR RESPONSE TO GRAZING

Plants of the prairie may be classified according to the manner in which they respond to grazing. Such lists have been compiled only after constant checking of the behavior of the species concerned in many pastures over a period of twelve years. The plants have been observed under various degrees of grazing and their normal abundance has been checked in adjacent prairies or in strips of native prairie along roadsides and outside the fences.

GRASSES THAT DECREASE UNDER GRAZING

In common usage, species in this group are designated as decreasers. The class includes all of those grasses that are best liked and most readily grazed by cattle. The most important grasses (all perennials) that tend to decrease in abundance more or less rapidly when true prairie is grazed, and may finally disappear, follow.

Andropogon gerardi	*Panicum virgatum*
Andropogon scoparius	*Stipa spartea*
Sorghastrum nutans	*Koeleria cristata*
Spartina pectinata	*Sporobolus heterolepis*
Elymus canadensis	*Sporobolus asper*

Big bluestem has high palatability; it not only is regularly grazed but also is selected by stock although other forage is abundant. When it is grazed, many new shoots are produced from the bases of stems and from shallow rhizomes. The shoots give rise to an abundance of leaves just above the soil's surface. Even when it is grazed to within 1 to 1.5 inches of the soil, considerable green tissue is left. Leafiness increases with grazing. This grass makes its greatest radial spread in spring, while using the food reserves of the previous summer. Hence early grazing inhibits normal development and decreases the growth of both the rhizomes and the new shoots.

Since the best stands on rolling land occur in and along ravines, grazing here is soon uniform and intense. With the increase in light consequent to the pasturing of the bluestem, bluegrass spreads rapidly between the stems and tufts of this taller grass. Both grasses are then grazed within an inch or two of the soil. No flower stocks of bluestem are formed; year by year, bluegrass increases, and finally the bluestem entirely disappears.

On uplands, big bluestem persists longer; the last remnants are found on steep slopes of rough land and on banks of deep ravines. Unless it is grazed in the early part of the growing season, it becomes less palatable and the woody stems are left during the late summer and fall if better forage is available. Even when it is grazed, the flower stalks of the isolated bunches may escape being eaten or trampled for a long time. During the following season, the old, dry leaves and flower stalks are conspicuous and are intermingled with those of the current year's growth. But when forage becomes less abundant, the green stems are sought, and even the cured leaves are not unpalatable. Finally, big bluestem disappears entirely.

Little bluestem is not as palatable as big bluestem. It is readily eaten in spring and early summer, but is often avoided after its woody stems are produced. Where it is moderately or closely grazed, it produces good forage throughout the entire growing season. Chemical analysis of plants in eastern Kansas indicate that ungrazed bluestems are very nutritious until July. After this, they become less leafy in proportion to the stems, and they gradually decrease in nutritive content and palatability (Aldous, 1938). They are valuable pasture plants because—if moisture is adequate—their excellent growth in June and

Fig. 99.—Deterioration of little bluestem under close grazing. Ungrazed bunch, about 28 inches tall (left), with its base surrounded by a new growth from peripheral tillers, which are grazed with part of the old stems. This is repeated until only a few stems remain. Invading bluegrass (right) replaced 75 percent of the overgrazed shoots of the bluestem.

July is supplemented by a considerable amount of forage in August. A moderately grazed bluestem pasture will provide palatable and nutritious forage throughout the growing season.

Probably because of its habit of growing in tufts and bunches, little bluestem does not withstand grazing as well as big bluestem; it disappears earlier, lingering longest on the least-grazed, roughest land (see Fig. 99). Experiments show that the greatest injury to both bluestems arises from close grazing at the beginning of the growing season. This is due to the fact that food reserves are mainly accumulated at this time. Hence pastures are greatly improved by protection or light grazing during May and early June. Later grazing is much less harmful. Deferred grazing, until after the seed is mature, results in a great waste of forage. In early stages of grazing, many bunches are left entirely unmolested, especially on steep hillsides of rolling uplands. On level or moderately sloping land where grazing has been uniform, all bunches have disappeared.

Needlegrass renews its growth very early in spring, it produces seed in late May or early June, and, in prairie, it remains more or less quiescent until fall, when it continues its growth until it is overtaken by freezing weather. It is palatable and nutritious. The bunches are grazed rather closely and are eaten repeatedly as new foliage appears. Consequently, there is very little reseeding, and the original stand, if grazed early in spring and late in fall (when other forage often is dry), disappears early. Information on other decreasers may be found in the original bulletin (Weaver and Hansen, 1941).

GRASSES THAT INCREASE UNDER GRAZING

Although much grazed, certain native grasses profit so much by the waning of the preceding species that they increase greatly. Ten of the most important species follow.

Poa pratensis	*Buchloe dactyloides*
Agropyron smithii	*Panicum scribnerianum*
Bouteloua curtipendula	*Panicum wilcoxianum*
Bouteloua gracilis	*Carex pennsylvanica*
Bouteloua hirsuta	*Agrostis hyemalis*

Poa pratensis, Kentucky bluegrass—contrary to the common belief—is probably not indigenous to the United States (Carrier and Bort, 1916), although—with the exception of timothy—it is the most important perennial grass that is cultivated in North America (U.S. Forest Service, 1937). Since the settlement of the midwest, it has successfully invaded true prairie, due to mowing and grazing, everywhere in the Missouri River Valley. A survey of 100 upland prairies in six midwestern states showed that bluegrass was of wide distribution and of considerable abundance, comprising about 5 percent of the basal cover of the vegetation. An even greater amount was found in low prairies, where it constituted about 9 percent. In this study it is treated with the native prairie grasses.

This bluegrass is a dense turf- and sod-former, producing large numbers of slender, shallow, creeping rhizomes. These give rise to tufts of grass above ground from which arise a profusion of fine, fibrous roots. It is a prolific seeder, and the tufts from the seedlings soon develop into a compact sod. It is the first grass to resume growth in spring, and it soon produces an abundance of nutritious forage which is highly palatable to all classes of stock. Bluegrass is not only unusually resistant to heavy grazing, it also is able to maintain its hold and actually to increase a stand even where the soil is much trampled and the grass is closely grazed. Its productivity decreases during the hot midsummer, but renewed growth begins in the fall and continues until freezing weather prevails. During dormancy it is very resistant to adverse conditions.

Early spring growth, when the soil usually is moist and most other species are dormant, greatly favors the spread of bluegrass. Its increase at first is very irregular and occurs in isolated patches. It is favored by close, selective grazing of big bluestem, and it makes its first appearance as a continuous sod in the bottoms of swales, along paths, and about gates. Although bluegrass is very palatable, aggressive, and persistent, it is rather low-yielding and not very drought resistant (see Fig. 100).

Side-oats grama increases slowly but consistently under grazing due to rhizomes and its prolific seeding habits. This occurs despite the fact that it is well liked by stock (because of its leafiness) and is closely cropped. In fact, it compares favorably in feeding qualities with the bluestems, being readily eaten when dry.

Blue grama is one of the most important grasses of the midwestern ranges, where it formerly occurred in mixed prairie as an understory to the native mid grasses. In true prairie, it occurred sparingly, and usually only in the driest situations, where the mid grasses did poorest and where this short grass was not entirely shaded out. In this region, this very palatable, drought-resistant grass forms a continuous sod—or only patches of sod interrupted by other grasses—depending upon the available soil moisture. It has the ability to remain dormant during drought periods, even those so severe as to kill the tops of the bluestems, and to become green and immediately resume growth when water is again available during summer or fall.

Fig. 100.—Effect of drought on bluegrass pastures in eastern Nebraska. This grass was killed on uplands, except in the shade of weeds, and finally almost disappeared elsewhere. Photo was taken Apr. 25, 1935.

Buffalo grass is one of the most important forage species of the short-grass ranges of the Great Plains. Like blue grama, it is of minor significance in native true prairie since it cannot endure the shading of its taller competitors. Under grazing, however, it occurs widely, especially over the arid western portion of the true prairie. Its rapid spread results from propagation by stolons which root readily where they come in contact with moist soil.

GRASS INVADERS

Certain grasses, mostly weeds, invade old pastures after many of the native grasses have died and left the cover broken. Light at the soil surface increases greatly when the tops of prairie plants are more or less completely removed by grazing. This, combined with the opening of the sod by the death of certain species, permits a host of weeds (none of which is found in virgin prairie) to become established. The change in the light relation is marked. Under a good stand of big bluestem, light frequently is reduced to 1 to 3 percent of full sunshine. In upland prairie, light intensities of only 15 to 20 percent have been regularly recorded. Moreover, with a reduced stand of grass, less water is absorbed by the native plants and more is left for the invaders.

The chief invaders include wire grass (*Aristida oligantha*), weedy brome grasses (*Bromus secalinus* and *B. tectorum*), crabgrass (*Digitaria sanguinalis*), stinkgrass (*Eragrostis cilianensis*), little barley (*Hordeum pusillum*), and several species of dropseed (*Sporobolus*). Nearly all are annuals of low forage value. Other species, such as Canada bluegrass (*Poa compressa*) and sand dropseed (*Sporobolus cryptandrus*), are perennials.

CLASSIFICATION OF FORBS, BASED UPON THEIR RESPONSE TO GRAZING

Prairie plants—other than grasses—that decrease under grazing constitute a long list of important species. All are perennials; in fact, more than 95 percent of all important true prairie species are long-lived. Among these, the legumes especially furnish much nutritious forage. They include the following:

Amorpha canescens (Lead plant)	*Lespedeza capitata* (Bush clover)
Astragalus canadensis (Canada milk vetch)	*Petalostemum candidum* (White prairie clover)
Astragalus crassicarpus (Ground plum)	*Petalostemum purpureum* (Purple prairie clover)
Desmodium canadense (Showy tick trefoil)	*Psoralea argophylla* (Silver-leaf psoralea)
Desmodium illinoiense (Illinois tick trefoil)	*Psoralea esculenta* (Prairie turnip)
Glycyrrhiza lepidota (Wild licorice)	*Psoralea tenuiflora* (Scurfpea)

The most abundant and widely distributed is lead plant, which annually may furnish 150 or more pounds of air-dry forage per acre. All of these legumes are readily eaten and are especially relished when they are young. Native sunflowers, several asters, prairie rose, prairie coneflower, blazing stars, and certain goldenrods more or less completely disappear under long-continued grazing. A list of 32 species is given and several of these are described in the midwestern pasture bulletin (Weaver and Hansen, 1941).

Many prairie forbs increase under grazing since they are either entirely uneaten by stock or are grazed so sparingly that they are not much handicapped. Among those that become bad weeds are yarrow (*Achillea millefolium*), many-flowered aster (*Aster ericoides*), prickly pear, ironweed (*Vernonia baldwini*), and certain goldenrods (*Solidago missouriensis* and *S. rigida*).

The leaves of ironweed are bitter and the plant is avoided by

Fig. 101.—Bluegrass pasture before the great drought almost ruined it by introducing invading weeds, chiefly ironweed (*Vernonia baldwini*).

stock. When the bluestems and other grasses about the plant are eaten, it rapidly increases its area by means of rhizomes and develops large, bushlike clumps 1 to 3 feet in diameter (Fig. 101). Many-flowered aster is a weed of much importance in uplands. It is not grazed, or is grazed only incidentally with other vegetation. It has an excellent system of rhizomes, thrives under grazing, and increases considerably. Only the young stems are mostly edible; the older ones become tough and woody. The tough, woody stems and thick coriaceous leaves of stiff goldenrod are not eaten by livestock. Seed is produced in abundance and new plants grow rapidly, especially in bare places. Thus, for one reason or another, forb increasers thrive as the pasture degrades from good to fair and then to poor condition.

WEEDS OF PASTURES

As long as a prairie is undisturbed except by mowing, weeds find no place in it; but when grazing and trampling open the cover, a host of these unwanted plants invade. At first they may find a home only in the trampled parts about gates and watering places, and from there spread into the pasture as bare spots occur.

Some of the most important species are annual ragweed (*Ambrosia elatior*), perennial ragweed (*A. psilostachya*), bull thistle (*Cirsium lanceolatum*), Canada thistle (*C. arvense*), wavy-leaved thistle (*C. undulatum*), gumweed (*Grindelia squarrosa*), Pursh's plantain (*Plantago purshii*), vervain (*Verbena stricta*), and spotted spurge (*Euphorbia maculata*).

The origins of native midwestern pastures were the grazed prairie. They were composed, at first, of grasses and forbs that decrease under grazing, and of another group of grasses and forbs that increase when the best forage is removed by grazing animals. These constitute our excellent and good grades of pasture. A fair type of pasture occurs where invasion of a third group of weedy grasses and weedy forbs begins. Unless good pasture management is maintained, degeneration into fair or even poor pastures may result. An examination of a typical native pasture will be made in Chapter 12 to ascertain the amount (percentage) of decreasers, increasers, and invaders that occur in eastern Nebraska.

When the tops of prairie plants are continuously and more or less completely removed by grazing, light at the soil surface increases greatly. This, combined with the opening of the sod by the death of certain species, permits a host of weeds, none of which is found in virgin prairie, to become established. The following are the more important species—most of them weeds—that are not a component part of true prairie vegetation and that regularly occur only in lower-grade pastures.

Aristida oligantha is an annual three-awn grass or wiregrass. Since it is not relished by grazing animals, the patches in which it occurs are frequently ungrazed. The abundant debris accumulates and may remain for a long time.

Bromus secalinus (and closely related varieties of chess) and *B. tectorum* (downy chess) have long occurred in moderate amounts in low-grade pastures, especially where the soil had been nearly bared. Since the drought, they have increased enormously and cover large areas, especially moist soil in low-grade pastures (see Fig. 102).

Eragrostis cilianensis, commonly called stinkgrass because of its odor, and *Digitaria sanguinalis* (crabgrass) are at first found only in trampled areas about gates, watering places, and in old paths where denudation is nearly complete. They then become much more plentiful and sometimes dominate very poor pastures.

Poa compressa, Canada bluegrass, was a frequent but rarely important component of pastures until after they had greatly deteriorated. The cover afforded by this species is open and its forage much less in quantity and lower in palatability than that of Kentucky bluegrass. Drought seems to have promoted its abundance where it was previously established.

Fig. 102.—Weedy pasture near Lincoln (top). The invaders are wavy-leaved thistle (*Cirsium undulatum*) and species of brome grass. A common weed, *Bromus commutatus* (bottom), that was especially common during the great drought.

Other grasses of this type are *Hordeum jubatum*, *H. pusillum*, *Panicum capillare*, *Schedonnardus paniculatus*, *Sporobolus neglectus*, and *S. vaginiflorus*. The annual ragweed is very widespread. It is sometimes scattered thinly throughout, but it is very dwarfed in dense sod. Since the plant has a very bitter taste, it is scarcely ever eaten. The perennial ragweed propagates by means of long, horizontal underground parts as well as by seed. It is found commonly, and often abundantly, in fair- or lower-grade pastures, particularly those that have much bluegrass. As the bluegrass sod becomes broken, stands of this unpalatable weed become more dense.

Three species of thistles are commonly found in weedy pastures. The bull thistle is an introduced weed. The tall plants, of dark-green color, are so well armed with stout prickles that they are avoided by stock. They frequently form such dense, thicketlike growths that they protect the grasses about them from grazing and the grass may thus produce seed even in well-grazed pastures. It is seldom found abundantly until the grass cover is much worn by grazing and bare places favorable to its growth appear. The smaller *Cirsium arvense*, or Canada thistle, is less common, but it is a much more troublesome weed. It spreads rapidly and widely by underground offshoots, and its wind-blown seeds give rise to seedlings in very distant places. Entire pastures are sometimes ruined by unchecked infestations of this very prickly, un-grazed, perennial weed. The wavy-leaved thistle is common to mixed prairie and on the western edge of true prairie, but it occurs rarely—except in pastures—eastward. It is a large, coarse, light-colored plant with abundant root offshoots that produce clusters of rosettes very early in spring.

The gumweed is a biennial. The first year, this plant forms a rosette about the size of a dandelion's. The next summer, the erect stems grow to a height of 2 to 4 feet, branch considerably, and produce an abun-dance of yellow-flowered heads. These are quite glutinous, which ac-counts for the common name. These inedible plants frequently form a dense, brushlike cover in old pastures where there is much broken sod.

Verbena, an erect vervain, is a characteristic weed of nearly all medium- or low-grade pastures. The woody stems grow rapidly and are 2.5 to 3 feet tall late in June, when the first blossoms appear. Blossom-ing continues until late fall. The small, individual blue flowers are clustered in erect spikes. This bitter-leaved plant is rarely eaten, even in very low-grade pastures where good forage is rare. When all the grasses are dry, this plant still remains green, and often blooms profusely. The preceding species also are examples of numerous other invading forbs.

As was stated in the textbook, *Plant Ecology* (Weaver and Clements, 1938):

> The degree of overgrazing is shown by two types of indicators—those due primarily to the fact that they are not eaten, and those that invade because of disturbance. The most palatable species are eaten down, thus rendering the uneaten ones more conspicuous. This quickly throws the advantage in competition to the side of the less palatable ones. Because of more water and light, their growth is greatly increased. They are enabled to store more food in their

propagative organs as well as to produce more seed. The grazed species are correspondingly handicapped in all these respects by the increase of less palatable species and the grasses are further weakened by trampling as stock wander about in search of food. Soon bare spots appear that are colonized by weeds or weedlike species. The weeds reproduce vigorously and sooner or later come to occupy most of the space between the fragments of the original vegetation. Before this condition is reached, usually the stock are forced to eat the less palatable species, and these begin to yield to the competition of the annuals. If grazing is sufficiently severe, these, too, may disappear unless they are woody, wholly unpalatable, or protected by spines.

Excellent and good pastures, because of the irregularity of grazing, usually have an uneven ragged appearance. Fair pastures appear smooth and pleasing in appearance, a result of more uniform grazing. Poor pastures are apt to appear rough and patchy, but in a different manner than those of high grade, because of their weedy nature.

Samples from pastures in early, medial, and late stages of degeneration showed consistent decreases in underground materials (Weaver and Harmon, 1935). The following data are taken from 25 half-square-meter samples that were taken to depths of 0 to 4 and 4 to 12 inches in various upland pastures. Differences sometimes were apparent after only two years of overgrazing. Decreases in the dry weight of underground plant parts were 35, 40, and 72 percent, respectively, from the original sod in the surface 4 inches. Decreases were 22, 23, and 69 percent in the 4- to 12-inch samples. Total decrease from the early to the late stage of degeneration was from 2.17 tons per acre to 0.95 ton per acre at 0 to 4 inches depth, and from 0.86 to 0.34 ton per acre at 4- to 12-inch depths.

XII.

Ecological Studies in a Midwestern Range

Native ranges and pastures are among the most important resources in the western half of the United States. Forage produced in native pastures is an important crop that is largely within the control of man. A study of the activities of grazing animals—as well as the vegetation—is essential to an understanding of proper range management. The purpose of this study was to ascertain the amount, composition, and consumption of forage, as well as the grazing activities and their effects upon the composition and distribution of the vegetation (Weaver and Tomanek, 1951).

There is continuing need for a better understanding of range vegetation and its use by grazing animals. Range preservation and improvement in the Midwest are largely matters of wise use and proper management of our natural grasslands. A more comprehensive understanding of the range itself is needed—of its forage, of soil and water supply, and of the factors that influence a proper distribution of livestock. The degree of the utilization of forage that will result in the maintenance of excellent or good range conditions and in the improvement of a range in fair or poor condition should be ascertained. Soil conservation on range lands is accomplished primarily by improving the vegetation.

The pasture we selected for study is almost completely surrounded by natural grassland. It is part of a long range of rolling hills northwest, west, and southwest of Lincoln, Nebraska, which is covered with thousands of acres of natural grassland. Many of the larger prairies and ranges are scarcely changed from their original condition. Rattlesnakes, coyotes, and various species of the original rodent population are common. The cattle are not subjected to herding or driving but remain undisturbed throughout the summer. This range was observed by the writer over a period of many years, and intensive studies in it were made over five years (1946–1950) (Weaver and Tomanek, 1951).

The experimental range is 5.5 miles north and 3 miles west of the University of Nebraska. It is 1 mile long north to south, and a half-mile in width. About 30 acres in the southwest corner have been

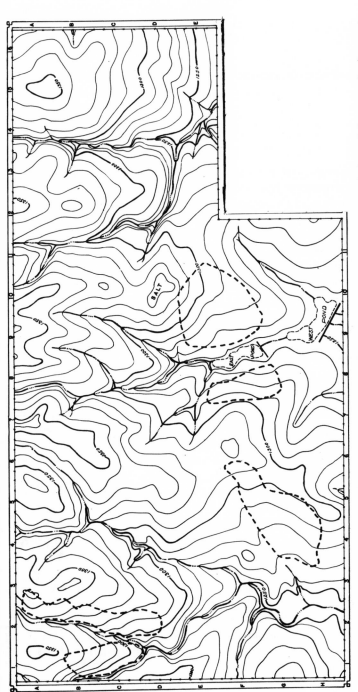

Fig. 103.—Topographic map of experimental pasture with contour intervals every 10 feet. The general drainage is from east (top) to west. Five deep ravines cross or nearly cross the pasture. There are 6 high hills or ridges that transect the area from east to west. Altitude is 1,350 feet on the east and 1,250 feet in the ravines and ponds on the west. Two areas near the ponds (enclosed by broken lines) were used for ascertaining yield and consumption of forage in fair range conditions. One area, farther north (left), was used for similar sampling in good range conditions, and two areas in the northeast corner were used for sampling on excellent range.

broken and cultivated, and the total area is approximately 290 acres. The topography is that of rolling upland. Hills, 50 to 100 feet high, are separated by deep ravines (see Fig. 103). These are dry except for a few days after rains. An exception occurs near the center of the pasture where underground seepage produces a small but often continuous stream of water. Two artificial ponds are located at the lower end of this ravine. Formerly, water was obtained at the farmyard, one-fourth mile south of the west pond. The occurrence of water in only one place resulted in a very unequal distribution of grazing. This is accentuated by the steepness of the hills, which often have slopes of 9 to 11 degrees and even-steeper banks of ravines must be crossed by the livestock.

Grazing began in this prairie in 1903, although hay was cut in autumn where the grazing had been light (usually in the north end) and where the topography permitted cutting. Originally, the vegetation was similar in all parts of the pasture, but the cover was less dense, of course, on the ridges. After several years it became clear that production in the south part was declining. This occurred because grazing was always closer in this area (adjacent to the farmyard) and because water and salt could be obtained. The pasture has never been partitioned by fences. The rate of stocking in the past 47 years had been very constant according to the owner. It has been about the same each year: 85 animal units (mostly cattle), including 14 horses. The rate of one animal unit for each 3.4 acres seems not to have been too great, but the distribution of the livestock was poor. Grazing began between April 15 and May 10 and ended late in October. The horses, however, had access to the pasture all winter. Since this heavy stocking was more or less concentrated in the southern two-thirds of the pasture, it could result only in overgrazing and deterioration of this part. In 1949, excellent range conditions were confined largely to the northern one-fourth of the prairie. Good conditions occurred over most of the north-central portion, but much of the remainder was in only fair range condition (see Fig. 104).

Although the loam soil is rich and deep and produces—under a mean annual precipitation of 27.8 inches—1 to 1.5 tons of hay per acre, the grazing was so concentrated in the southern two-thirds of the range that the native bluestems and accompanying prairie grasses have been replaced largely by Kentucky bluegrass and blue grama.

After long and careful observations, the three range-condition classes were delimited and marked out on a specially constructed topographic map. The basis for mapping was entirely that of the percentage composition of the vegetation.

Fig. 104.—View in central portion of the experimental range. The far-distant livestock are on the central ridge, a quarter-mile southward. Excellent range conditions are shown in the foreground, where bluestems are abundant.

Excellent range, located largely in the northern third of the pasture, consisted mostly of grasses which decrease under heavy grazing—chiefly big bluestem and little bluestem. An abundance of old, ungrazed bunches of little bluestem and tall dropseed, together with very little bluegrass and much debris, characterized this class, which composed 13 percent of the range (see Fig. 105).

In the good range-condition class, which occupies 30 percent of the pasture, the old, standing bunches of the preceding decreasers were not very continuous. There was much less debris or mulch. All the vegetation had been grazed at some time, moderately in places but more closely in others. The increaser, Kentucky bluegrass, had spread in large amounts, and—with the preceding characteristics—it clearly demarked good from excellent range, most of which occurred in the north half or southeastern part of the pasture at a considerable distance from water.

Grazing in spots or patches in the good pasture class was common (see Fig. 106). It was largely by this means that the area was opened to

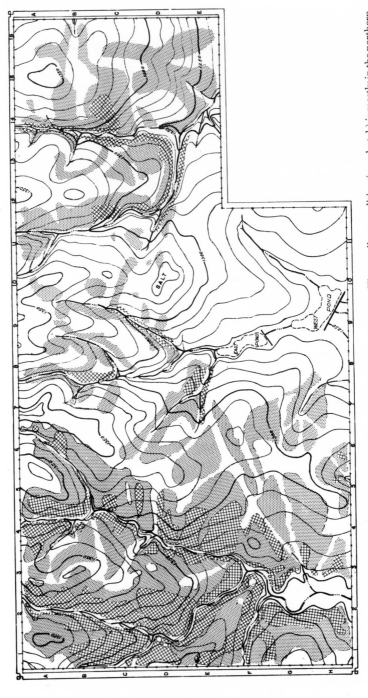

FIG. 105.—Range-condition classes are superimposed upon the topographic map. The excellent condition (crosshatch) is mostly in the northern portion of the pasture; otherwise, it nearly always occurs on steep banks that border ravines. The unshaded areas, the most extensive areas in the central and southern portions, represent the fair range-condition class. The dotted areas represent vegetation in good range-condition class.

Fig. 106.—An example of spot grazing in good range-condition class. Bluegrass rapidly increased in these closely grazed spots and patches. About half of this class of range consisted of such places which were often connected and formed a network in which close grazing regularly occurred.

final, general grazing. Once a bit of vegetation a few square feet to a few square yards had been freed from the old growth, it was grazed again and again and was a center for the establishment of bluegrass. Such areas occur as spots or patches, but under continued use they are constantly enlarged. Later, they merge to form a network of closely to moderately grazed places among ungrazed bunches and debris-filled vegetation. Finally, the grazed areas predominate in extent over the roughs and relict bunches which occur only here and there. These also may be consumed in periods of drought or overstocking, but often—before this happens—another type of grazing usually has occurred. This consists in first grazing the tops of big bluestem, prairie dropseed, and other plants, but also of little bluestem and tall dropseed as well. The effect was to reduce the general level of the vegetation closer and closer to the soil. The decreased competition by most prairie grasses for light and water was very favorable for the spreading of bluegrass, side-oats grama, and blue grama, and for the invasion of weedy grasses and forbs. These increasers constituted practically half of the vegetation. Such areas were usually closely or at least moderately

Fig. 107.—Fair range-condition class north of the upper pond. The bunches are mostly little bluestem, but some are tall dropseed (*Sporobolus asper*).

grazed. Continued overgrazing in the good pasture reduces it finally to the fair range-condition class.

The fair range class, which occupied about 57 percent of the pasture, had the appearance everywhere of having been closely grazed (Fig. 107). Only small areas of fair range occurred in the northern third of the range. Increasers, such as blue grama and bluegrass, composed most of the vegetation. There were many invading grasses, such as hairy chess and little barley, but bluestems and other decreasers were few.

Only about 3 percent had been reduced by overuse, trampling, etc., to the poor range-condition class. These were places in and about paths, near the ponds and salting area, and under the shade of the few trees which were present.

The patches of wolfberry (*Symphoricarpos occidentalis*) are not included in this study. Although they are most abundant in overgrazed places, they occurred nearly throughout the pasture and occupied 3 to 5 percent of the land. Such areas furnish practically no food from either grass or shrub.

SELECTION OF STUDY AREAS

Three areas, ranging from 7 to 10 acres in extent, were selected in which to make detailed, quantitative studies (see Fig. 103). These selections were based on the concepts of excellent, good, and fair grades of pasture or range-condition classes as defined by Weaver and Hansen (1941).

Selection of an area that was representative of excellent range conditions was not difficult—not even in spring, before growth was resumed—because of the abundance and character of the previous year's vegetation that remained on the soil. Flower stalks of little bluestem were abundant; much high-grazed big bluestem remained; the dried, curled leaves of side-oats grama were conspicuous; and there was a mulch of lodged plants or plant parts to a depth of 1 to 4 inches. Other evidence of an excellent range-condition class was the prevalence of many of last year's prairie forbs, even the most palatable ones, such as lead plant (*Amorpha canescens*), many-flowered psoralea (*Psoralea floribunda*), silvery psoralea (*P. argophylla*), and many others.

The sampling area in the good range-condition class was chosen for its apparent uniformity of vegetation, degree of slope (4 to 7 percent), and northern exposure. Here it was evident that continued grazing had been sufficiently heavy to weaken or destroy much of the native vegetation and to permit a great spreading of bluegrass. This, of course, greatly increased the basal area. In excellent range, the total basal area of vegetation was 18 percent, but in good range this was 29.3 percent. It will be shown that bluegrass increased still further in fair range, where blue grama also was plentiful. Here the total basal area was 38.8 percent. Patches of vegetation that were several square yards in extent and contained mostly bluegrass occurred throughout the good range class. Other, smaller, closely grazed places were numerous. Conversely, many places where little bluestem had been ungrazed, or at least grazed high, were scattered about. These often bore the flower stalks and rank foliage of the preceding year.

The larger portion of the central and southern parts of the pasture was representative of the fair range-condition class. Bluegrass, blue grama, side-oats grama, sand dropseed, and may weedy forbs composed most of the vegetation. The short-lived hairy chess and little barley (*Hordeum pusillum*) often were abundant. Throughout a long period of years, grazing had usually been so close that relatively little debris was left on the soil. Paths occurred throughout in great numbers, most of them converging near the ponds or at the adjacent salting site on the hill.

Composition of the Vegetation

Vegetation in each range-condition class in the selected areas was sampled in numerous well-distributed, square-foot plots. Lines were marked out in such a manner that they crossed the entire sampling area lengthwise in at least two different places. Sampling along the lines was at random, usually at intervals of 12 paces. Of the 150 samples taken in each range-condition class, 50 were obtained in June, 50 in July, and 50 in August. All of the early samplings were made over the entire area, but the later samples were taken between those that had been obtained earlier.

The quadrat frame consisted of a strip of steel and was constructed to enclose three sides of a square foot. The open side permitted the frame to be slid on the soil and through the vegetation of the selected area. The fourth side was then placed across the open end and fitted into slots to hold it in place. Location of the sample was selected at random. After the placing of the quadrat frame, the square foot was divided into four equal areas by laying long steel quadrat pins across the frame in two directions. The pins rested in shallow grooves that were made by filing the top of the frame. The basal area occupied by the vegetation 1.5 inches above the soil surface was estimated in each part of the quadrat and the average of the four numbers was recorded.

Percentage composition of the vegetation was next ascertained. In doing this, the total vegetation—regardless of its amount—was considered as a unity, or 100 percent. Usually, two or three grasses formed the bulk of the vegetation, the total of their separate percentages of composition amounting to 80–90 percent. The percentage for each species was based upon the part of the total basal area (now considered as 100 percent) that each furnished. The basal area of forbs was usually very small, and was not recorded unless it totaled 5 percent. Likewise, the basal area of every grass or sedge was recorded only if it furnished at least 5 percent of the total basal area. Thus the species found in small amounts were ignored and the 100 percent basal area of the vegetation was divided among the abundant species.

The percentage composition in each of the three range-condition classes is shown in Table 11. Even casual examination shows that the composition is distinctly different in each range-condition class. The decreasers compose 81 percent in the excellent class, but their amount was reduced to 47 percent in the good class, and to only 7 percent in the fair class. The bluestems furnish the bulk of this type of vegetation in each class: 72 percent in the first, 45 percent in the second, but only 7 percent in the third class.

The rapid gains of increasers—from 18 to 50 percent and then to 66 percent—closely paralleled losses by decreasers, except in the fair grade, where a large percentage (21 percent) of invaders occurred. Bluegrass was the most abundant increaser. Although it is highly

TABLE 11
PERCENTAGE COMPOSITION OF THE VEGETATION IN EACH RANGE-
CONDITION CLASS. FIGURES ARE THE AVERAGES OF 150 SAMPLES
TAKEN DURING THE SUMMER

Species	Excellent	Good	Fair
Decreasers			
Big bluestem	32.7	17.3	5.9
Little bluestem	39.0	28.1	1.0
Tall dropseed	5.4	0.5	. . .
Prairie dropseed	2.4	0.6	. . .
Indian grass	1.4	0.4	. . .
Needlegrass	0.2	. . .	. . .
Subtotal	81.1	46.9	6.9
Increasers			
Kentucky bluegrass . . .	11.4	28.4	36.3
Purple lovegrass	2.4	14.8	3.0
Side-oats grama	1.8	4.9	8.1
Blue grama	. . .	. . .	13.8
Sedges	1.8	1.3	3.6
Scribner's & Wilcox's panic grass .	0.4	0.3	0.3
Buffalo grass	. . .	. . .	0.6
Hairy grama	. . .	. . .	0.1
Subtotal	17.8	49.7	65.8
Invaders			
Sand dropseed	0.1	0.7	12.1
Beadgrass	0.1	0.4	3.4
Hairy chess	. . .	0.2	2.2
Little barley	. . .	. . .	2.9
Prairie three-awn . . .	. . .	. . .	0.2
Western wheatgrass . . .	. . .	. . .	0.1
Tumblegrass	. . .	. . .	0.1
Subtotal	0.2	1.3	21.0
Forbs	0.9	2.1	6.3
Grand Total . . .	100	100	100

palatable, it withstands grazing well and is persistent and aggressive. Moderate-to-close grazing of the prairie grasses and weather conditions that are favorable to the growth of bluegrass result in its occupation of more and more of a pastured prairie.

Invading grasses were only rarely sufficiently abundant to compose 5 percent of the vegetation in even one sampling area. Sand dropseed, however, composed 12 percent in the fair class, and six other species were present in smaller amounts. The total for invading grasses was 21 percent. Forbs, too (mostly western ragweed and other invaders), were by far the most abundant (6.3 percent) in this class.

Hundreds of samples were taken at a random 8 paces apart and at a distance of only 9 feet on both sides of the boundaries between the range classes. Comparison of these samples within a total distance of 1.25 miles showed a remarkable similarity in percentage composition to the samples that were taken in the experimental area in each class of range.

COMPARISON WITH PREVIOUS STUDIES

Change in the composition of vegetation on a range is ordinarily a gradual process. Vegetation in one range-condition class degenerates under poor management into the next lower class. Under protection or proper usage, vegetation that denotes a higher range condition may develop. Since both degeneration and regeneration occur gradually, the resulting good pastures are not all equally good, nor are the excellent ranges all of the same degree of excellence. There is a considerable degree of variation in every range-condition class. With continued observation and study, one can soon recognize high, intermediate, fair, and low types of good pastures.

In a previous research (Voigt and Weaver, 1951), twelve pastures —three in each of four range-condition classes—were selected for study. The result of this research afforded a concise basis for comparing data from the present findings; hence we will digress briefly to examine the results.

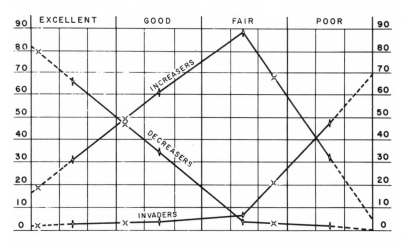

Fig. 108.—The probable trends of grass decreasers, increasers, and invaders between actual points determined by averaging the percentages from each of 3 pastures in each range-condition class. Projections of these lines (which are broken) show the probable relation to climax prairie on the extreme left and the low type of poor pasture on the right. The fair range-condition in this study, according to the percentage of each class of grasses (circles and *X* marks), is in the low type. But high types of both good and excellent range-condition classes prevail.

Grass decreasers, as a group, furnished 66.6 percent of the vegetation in the excellent pastures, 34.2 percent in good ones, 3.9 percent in those of fair condition, but only 1.9 percent in poor ones (Fig. 108). Total increasers averaged 30.5, 60.6, 88.0, and 31.7 percent in the several classes of pastures.

Invaders were few in excellent pastures (2.1 percent) and in good ones (3.1 percent). They increased to 6.2 percent in fair pastures, and were extremely abundant—47.5 percent—in poor ones.

The fair range condition in our one-pasture study, according to the percentage of each class of grasses (X) is in the low type. But a high type of both good and excellent range-condition classes prevailed.

YIELD AND CONSUMPTION OF FORAGE

Monthly and annual yields and consumption of forage were ascertained by means of 29-square-foot movable exclosures and by grazed control areas of the same size. Ten of these were used in the experimental area in each range-condition class. By selective clipping, the yield and consumption of decreasers, increasers, and invaders among grasses and forbs were ascertained separately.

Yield and consumption per acre (May to October, 1950) were 0.89 and 0.77 tons in fair range, 1.92 and 1.20 tons in good range, and 1.53 and 0.86 tons in excellent range, where there was little bluegrass and excessive mulch. The fair range was constantly and greatly over-grazed to secure 0.77 ton of forage per acre; the good pasture furnished 1.20 tons without overuse, and left a third of the yield as a protective mulch on the soil. Also, more forage was obtained from the lightly grazed excellent range (where the cattle selected only the very best) than from the fair range. Yield and consumption usually were greater during the wetter year 1949, but even less forage was consumed in excellent range.

GRAZING PREFERENCE OF CATTLE

Grazing preference refers to the taste an animal displays for any plant. It is measured by the choice of an animal if it is given free access to various species of plants. When a prairie is first grazed, the process is highly selective, and the Lincoln area vegetation contains about 250 species of plants per square mile (Steiger, 1930). Nearly all species are edible. Most of these, however, occur only in small numbers; many are even rare. Among the common and abundant grasses, sedges, and forbs, certain ones are greatly preferred by the cattle, others are less sought, and some are eaten only because they are intermixed with the best-liked species. Thus selective grazing occurs in all range-condition classes, at least where the amount of forage is plentiful.

Sampling for grazing preference was confined to the areas in each range-condition class where yield and consumption were measured. The sampling was done at random along numerous predetermined lines. The area of each sample was that of a steel ring 3 inches in diameter. Samples were taken along lines 5 to 10 paces apart. Twelve hundred samples were examined in May, and 1,500 samples during each of the three following months.

Grazing preference was highest for big bluestem in all range classes. Preference was equally high for little bluestem in fair range, but decreased greatly in good pasture, and even more in excellent range. Preference for bluegrass increased from the excellent to the fair range. Preference for six other grasses also was ascertained.

ACTIVITIES OF THE CATTLE

For a fuller understanding of the condition of the pasture and the causes of the distribution of the several range classes, it seemed necessary that one should understand the activities of the cattle as they grazed and traveled on the range. Only in this way can one clearly observe what they do, and, insofar as possible, ascertain why they do it.

Activities of the cattle were studied by continuously watching them and making records of their behavior at intervals of 15 minutes by day and 30 minutes by night. The observer was careful not to approach close enough to the herd to disturb any individual animal or to affect its activities. The cattle paid scant attention to anyone who worked in the pasture more or less regularly, but a stranger or a dog caused considerable disturbance, especially when the calves were small. The instinct to protect them was clearly shown not only by the mother but by every member of the herd.

Observations were not difficult on the rolling terrain, where one could observe the cattle clearly through binoculars. The studies were made during eight separate days between 4:00 a.m. and 9:00 p.m., and for five nights between dark and dawn. Unless the night was cloudless and the moonlight bright, a flashlight was employed. In addition, 16 half-days were spent in observation, and isolated observations were recorded almost every day during the grazing season.

The herd arose at daylight (about 4:30 a.m., or within an hour or more thereafter) and all the cattle were grazing by 6:00 a.m. Rising was much more prompt on a hot day but considerably delayed on a cool one. About two hours (7:00 to 9:00 a.m.) were spent drinking, licking salt, and standing about the ponds. On a cool day, drinking was postponed until late afternoon. The time was spent in resting. On a hot day, the cattle remained in or near the ponds until about 4:00 p.m.; then they grazed until dark.

Grazing on a typical day was resumed about 9:00 a.m., and, except for a short period for rest, continued until noon to 12:30 p.m. Many cattle rested until about 4:00 p.m., when grazing was resumed as they traveled slowly toward a bedding ground. About two-thirds of the cattle visited the ponds once again, but immediately returned, and all had bedded down by dark. The herd did not lie down at dark in any part of the range but only in preferred places. For the most part, three different bedding grounds were used. All of these were in excellent or good range classes and all were one-fourth to one-half mile from the ponds. A part of the herd grazed an hour or two (11:00 p.m. until 1:00 a.m.) at night. Grazing was extended over a greater period on a hot night.

The major part of the grazing occurred between 4:00 and 8:00 a.m. and between 5:00 and 9:00 p.m., when 67 and 71 percent of the herd, respectively, were thus engaged. About 10.5 hours were spent in grazing, nearly 5 hours in standing or traveling, and 8.5 hours lying down.

Cattle almost always grazed into the wind. There was a close correlation between the location of the cattle on any day and the direction from which the wind was blowing. The wind blew from a southerly direction on 48 percent of the days in the summer of 1950, and from a southerly direction—mostly southeast—45 percent of the time during a five-year period. The conclusion that the direction of the wind greatly affected the use of various portions of the range is confirmed by Allred (1950), who stated that domestic animals in the Southwest have the habit of grazing into the breeze. As a result, the south side of a pasture is generally overused.

Location of water and salt, as well as direction of wind, greatly affected the amount of grazing and trampling in different parts of the range. In general, both became less as the distance from water increased. Much traveling also occurred to and from the selected bedding grounds.

Movement of cattle through the pasture did not occur uniformly but nearly always either along paths or on much wider linear strips that were designated as grazing routes. These were generally found to be the easiest routes of travel—along ravines, crests of ridges, etc. Aside from topography and choice of forage, location of the ponds and the salting place in the same area over a long period of years has resulted in a well-used network of paths and grazing routes, which have been mapped.

Factors affecting the distribution of the livestock and the consequent overuse of some areas and underuse of others are discussed. Methods for better utilization of the forage are suggested. The resulting increase in forage and in gains of cattle would (in 1951) have quickly repayed the investment in fencing required for proper distribution (Weaver and Tomanek, 1951).

Since this study was completed, a pond has been constructed in the north end of the pasture. This has resulted in a great change in the use of the vegetation.

RÉSUMÉ

Degeneration from one class of pasture to the next-lower class is ordinarily accomplished slowly over a period of years.

> The more palatable species are eaten down, thus rendering the uneaten ones more conspicuous. This quickly throws the advantage in competition to the side of the latter. Because of more water and light, their growth is greatly increased. They are enabled to store more food in their propagative organs as well as to produce more seed. The grazed species are correspondingly handicapped in all these respects by the increase of less palatable species and the grasses are further weakened by trampling as stock wanders about in search of food. Soon bare spots appear that are colonized by weeds or weedlike species. The weeds reproduce vigorously and sooner or later come to occupy most of the space between the fragments of the original vegetation. Before this condition is reached, usually the stock are forced to eat less palatable species, and these begin to yield to the competition of annuals. If grazing is sufficiently severe, these, too, may disappear unless they are woody, wholly unpalatable, or protected by spines (Weaver and Clements, 1938, p. 470).

A succinct account of the degeneration of lowland prairie near Lincoln, Nebraska, under excessive grazing may be found in the writer's *North American Prairie* (1954a). The original account, "Changes in Vegetation and Production of Forage Resulting from Grazing Lowland Prairie," was published in *Ecology* (Weaver and Darland, 1948).

The degeneration of productive grassland has a profound effect upon the economy and welfare of the community. Once the early symptoms of deterioration are generally recognized, corrective measures may be taken to stop the downward trend and to improve the range. The concept of four different grades of pasture or four distinctly different range-condition classes has been extremely valuable to students of ecology, range examiners, and graziers. It presents a fixed concept of a pasture class or range condition with which another pasture may be compared. The placing of this concept on a percentage basis adds to its validity. It focuses attention more directly on each type of vegetation. Such a series of pastures is invaluable in teaching ecology and in an educational program of soil conservation. Cover of some sort is the chief tool used in conservation practice. This scheme of classification of range lands, with modifications, is now in general use by the U.S. Soil Conservation Service.

So thorough has been the destruction of the original prairie by plowing, overgrazing, and trampling by cattle that even small tracts of typical native grassland are rare in both the true and the mixed prairies of the Great Plains.

XIII.
Removal of Tops and Development
of Grasses

The persistence of remnants of bunches or mats of grasses in prairie that has been pastured for many years is of interest, as is also their gradual but slow rate of recovery under protection from grazing. The first summer, only a poor growth of scattered foliage of the blue-stems outlines—in a fragmentary manner—the location of the under-ground parts that have been thoroughly weakened by continuous depletion of their food supplies. A second growing season shows a marked filling in of the sodded areas or clumps and about half the normal production of foliage. During a third summer, the area is further extended, production of foliage is normal, and at least a few flower stalks and some seed are produced. It was further observed that the weakened grasses were more susceptible to wilting than similar species in adjacent prairie. The evidence pointed directly to a meager or inefficient root system.

Effects of Removal of Tops on Underground Parts

The effect of the removal of the photosynthetic area upon the growth of the tops has received considerable study. It has been repeatedly shown that the yield and vigor of the vegetation varies inversely with the frequency of the removal of tops. Less work has been done to ascertain the deleterious effects on root development of clipping or grazing the tops. Much of the early work was done with plants that were grown in water cultures.

It was decided to select representative samples of several of the most important grasses, to cut blocks of soil that contained them, and to transplant the blocks into large containers filled with soil that was free of roots and debris. One lot of each grass was to be clipped at 14-day intervals and the duplicates were to be grown as controls. Beginning on June 28, blocks of well-established sod of seven impor-tant native pasture grasses were transplanted into large containers, were grown out-of-doors, and were clipped at 2-week intervals (Biswell and Weaver, 1933).

The growth of tops, which were removed upon transplanting, was renewed immediately, as well as after each subsequent clipping between July and October. The foliage produced at each clipping was oven dried and its weight was ascertained. The dry weight of tops of big bluestem, little bluestem, switchgrass, blue grama, and buffalo grass increased for the first 3 to 5 intervals after the initial clipping, after which it decreased rapidly. The yield of Kentucky bluegrass increased after each clipping.

The total dry weight of tops from the clipped sods ranged from 13.1 percent (switchgrass) to 47.5 percent (side-oats grama) of the weight of the same species that were unclipped after transplanting. In buffalo grass, where the stolons were permitted to grow, it was 63.1 percent. The stands of weakened grasses in all containers were considerably thinned. Plants failed to produce new rhizomes and many of the old ones died.

The relative production of roots was more greatly reduced than that of tops, and the length of roots was greatly decreased. By volume, the reduction ranged from 3.5 percent of the controls (blue grama) to 18.6 percent (Kentucky bluegrass), although that of buffalo grass was 36.4 percent. Dry weight varied from 2.6 percent of that of the unclipped control in switchgrass to 20.6 in bluegrass. The average root volume and the average root weight of all species was 11.7 and 10.1 percent of that of the controls.

The roots of clipped grasses were of smaller diameter than those of the controls, and thus they were less efficient for the conduction of water or storage of nutrients. Plants that were weakened by repeated clipping renewed their growth slowly—if at all—after the soil had been frozen for several weeks.

It seems almost certain that, after transplanting, the production of new roots was delayed until the tops were well established. This was indicated by the delay in root growth of the species that were transplanted into the greenhouse in winter, and this was verified by experimental transplants the following summer. Thus within the clipped block of sod there appears to be competition for the accumulated food for building new roots and new shoots (see Fig. 109).

The common practice of pioneer farmers was to pasture the prairie for a year or more before they broke the sod so that it would be less dense when it was broken and the land could be more easily tilled. Grazing is a more-or-less-destructive process since it periodically removes much of the photosynthetic area of plants. An abrupt decrease in the amount of green tissue causes a corresponding decrease in the

FIG. 109.—Growth of tops and roots from similar vigorous bunches of little blue-stem 42 days after transplanting the sod blocks. Less than half of the root system is exposed. The control bunch (left) had a total of 150 roots that extended below the sod, but the plants that were clipped four times (right) had only 3 roots.

growth of the roots. Continued defoliation results in the destruction of the root system, and this is followed by death. Hence, unless reasonable precautions are taken, the effects of grazing are likely to become cumulative and cause serious deterioration of the range.

In another study, by Robertson (1933), on seedling grasses, clipping was found to be even more harmful than in the preceding study because the food reserves are indeed small for a time. Experiments with six prairie and pasture grasses showed that the removal of the tops by clipping had an immediately injurious effect that was measurable both above and below ground. The extent of injury depended largely upon the nature of the species and the frequency of the treatment.

Continuously overgrazed native pastures reveal several distinct stages of degeneration. Each stage is indicated by the dominance of certain grasses and forbs. Samples from pastures in early, medial, and late stages of degeneration showed consistent decreases in underground plant materials. Numerous half-square-meter samples were secured from several such pastures. On lowlands, decreases of 28, 40, and 77 percent in dry weight in the surface 4 inches occurred in the three stages of degeneration—decreases from the original prairie cover. On uplands, dry weight decreased 35, 40, and 72 percent. Similar losses were sustained in the 4- to 12- inch layer of soil. These data are taken from a bulletin by Weaver and Harmon (1935), in which the plant cover is also discussed in relation to rainfall interception, decrease in runoff water, and promotion of absorption. The bulletin also discusses underground plant parts in relation to porespace, soil structure, and erosion.

METHOD OF TESTING PLANT VIGOR

The degeneration of excellent or good native pastures and ranges into fair or poor ones is always preceded by a decrease in vigor of the most nutritious and best-liked grasses. These are nearly always the climax species. Decreased vigor may result from overgrazing or from drought. If this sign of range deterioration is observed and the stocking rate is decreased, or if grazing is deferred, or if the pasture is completely rested, the range will usually recover, and often will improve.

Vigor of vegetation, composition, and density are the most important indicators among plants of the condition of the range. Vigor is commonly shown by the size of bunches or clumps, and especially by the number of stems, the absence or presence of dead centers, and the partial or complete death of tufts and bunches.

An excellent test of vigor under conditions that are favorable for development is that of prompt renewal of growth in spring, or after grazing, or after transplanting. The last test permits exact measurements of heights, production of forage, and the rate of development and amount of new roots. Results are clearly shown for two species in

Fig. 110.—Development of tops and roots of similar blocks of sod of moderately grazed (left) and closely grazed (right) sand dropseed 28 days after transplanting the sods (on June 10).

FIG. 111.—Plants produced from closely grazed sod of buffalo grass (right) and ungrazed sod (left) during a period of 32 days. Yield from the ungrazed sod is more than twice as great. Also, there were twice as many stolons, with a total length 4 times as great, as those on the ungrazed grass.

Figures 110 and 111, where the boxes were 10 × 10 inches square and 24 inches deep (Weaver and Darland, 1947).

REGENERATION OF NATIVE PASTURES UNDER COMPLETE PROTECTION

The role of sand dropseed (*Sporobolus cryptandrus*), an early invader from the Great Plains during the drought of 1934–1940, was so important in the following study that a preliminary statement seems necessary, since this species is not common to true prairie.

Sand dropseed is a tufted perennial of xeric habit. In general, the tufts are small, but the larger bunches may have a basal diameter of 5 to 6 inches and 30 to 50 leafy stems, which form an open crown (Fig. 112). The pithy, solid stems are mostly erect, but some usually spread outward at various angles or may even grow parallel to the soil surface.

Fig. 112.—Development of sand dropseed (*Sporobolus cryptandrus*) in bared area (left) and under competition with bluestems (right).

Even vertical stems often spread outward at the base. The leaves are usually 5 to 12 inches long, with conspicuous, long, white tufts of hairs where the blade joins the sheath. Nearly all are confined to the lower half of the mature plant. Growth in Nebraska is renewed in spring long after that of bluegrass, needlegrass, Junegrass, and other species of northern extraction, but earlier than that of bluestems.

Growth is fairly rapid and production of flower stalks begins about June 15. During favorable seasons, the flower stalks are borne in great profusion and attain a height of 3 feet. During dry years, however, they may be reduced to one-third the normal size. The pale or leaden-to-purplish panicles are partly enclosed in the topmost leaf sheaths. The unenclosed portion spreads somewhat, but the panicle usually is narrow. The larger ones are often 15 inches long. Blossoming begins late in June and may continue until October.

The seeds, which mature in late summer or early fall, are produced in enormous numbers. Thousands of mature seeds have been obtained from a single enclosed panicle. The small seeds gradually fall out of the enclosing leaf sheath and may readily be carried along the ground by high winds for considerable distances. Germination is high but some seeds frequently lie dormant for many years. Tillers appear only a few weeks after germination, and the seedling soon becomes a small tuft or bunch of leafy stems. Thereafter, growth of ungrazed plants is even more rapid. Several flower stalks are produced and the seeds ripen, thus completing the round of life in a single year.

The species' resistance to drought is due in part to an excellent root system. This consists of a vast network of fine roots and masses of finely branched rootlets. The soil beneath the plant and for 8 to 12 inches on all sides is filled with a dense mat of roots to a depth of about 4 feet. This absorbing system is remarkably efficient.

The palatability of this grass varies greatly. In eastern Nebraska, cattle and horses graze it readily, often in preference to western wheatgrass. Not only are the leaves eaten, but also the stems and flower stalks—at all stages of development. The plants are often grazed closely, even where forage is abundant. Many of the former bluegrass and little bluestem pastures of eastern Nebraska became populated with a good cover of sand dropseed. Others were so greatly abused during the long drought that the once-fine stand of dropseed became represented only by dead crowns and by a few, small, weak survivors of this once abundant grass (Weaver and Hansen, 1939).

The study of plant succession has contributed more than any other single line of investigation to a deeper insight into the nature of vegetation. This is a study of the nature and rate of regeneration of a pasture that adjoined a tract of true prairie one-half mile long and one-quarter mile wide—of which it originally was a part (Weaver and Hansen, 1941a). The prairie and pasture covered gently rolling to moderately hilly land. The prairie had been mowed once annually for hay. A considerable portion of the prairie had been fenced and grazed for 23 years. This native pasture apparently had never been greatly over-grazed. It was examined in 1932 and described as being in good condition, with about half-dominance of bluestems and other prairie grasses and half-dominance of bluegrass. The cover was intact, weedy species were few, and only a modicum of the usual perennial pasture weeds occurred.

The pasture area enclosed for study was 4.5 rods wide and 31 rods long; it extended from the top of a low hill to (and including) its nearly level base. The slope varied from 1 to 5 percent. The adjacent prairie varied from the little bluestem type on the upper part of this north hillside to the big bluestem type on the lower slope, and included level but well-drained land. On the upper slope the silty clay loam of the A horizon was 10 to 12 inches deep, and the silt loam of the lower land had an A horizon that was 20 inches deep. Both soil types have deep, permeable B and C horizons. The average annual precipitation is 28 inches.

In 1937, owing to the severe drought of 1934–1936, both little bluestem and big bluestem occurred very sparingly. Bluegrass remained only in scattered patches, ranging from a few square decimeters to a square meter in area. Sand dropseed varied in occurrence from sparse to abundant, as did side-oats grama. Small amounts of blue grama, Junegrass, and a few other native grasses were found. Dense patches of peppergrass (*Lepidium densiflorum*) occurred throughout much of the pasture, and Pursh's plantain (*Plantago purshii*) was very abundant in widely distributed patches. These, with smooth goldenrod (*Solidago missouriensis*), many-flowered aster (*Aster multiflorus*), and horseweed (*Leptilon canadense*), were the major constituents of the weedy flora.

Five different lots of permanent square-meter quadrats were established in 1937, in which the increase or decrease of each species was determined quantitatively by the stem-count method. Ten of these were marked out at random in widely distributed areas where sand dropseed dominated; 10 were similarly located in local areas that revealed remnants of little bluestem; and 10 others were placed where relict patches of bluegrass persisted. A fourth lot of 10 was scattered widely to include small patches of blue grama, and three quadrats were established in representative bared areas that were occupied by peppergrass.

This method, though slow and laborious, was selected as the best method for studying the fragments of drought-damaged plants and the scattered individual invaders. Succession in grassland can be followed only by exact methods of counting, measuring, mapping, and comparing the vegetation year after year. In addition to the changes recorded in the quadrats, the more apparent variations in the vegetation in the entire protected pasture were studied throughout each summer. Thirteen species of grasses and sedges, 19 species of native forbs, and 13 ruderals composed the total vegetation in the 43 quadrats in 1937. This number, 45, increased to 58 four years later.

Drought occurred at certain periods during each growing season, but 1937, 1939, and 1940 were especially dry. Conditions for growth were very favorable in the spring and early summer of 1938.

Soil samples were taken at frequent intervals to depths of 4 to 6 feet in both pasture and adjacent prairie. They showed that the vegetation depended chiefly upon the current rainfall for its water for growth, since only small amounts of water occurred in the deeper soil layers at the beginning of the study. The water available for growth frequently was exhausted in the first and second foot. This condition, often combined with long periods of unusually high temperatures and extremely high rates of evaporation, retarded the normal development of vegetation. A summary of succession in the quadrats follows.

TABLE 12

PERCENTAGE COMPOSITION OF GRASSES AND SEDGES IN 1937 AND 1940, BASED UPON THE NUMBER OF STEMS

Species	1937	1940	Species	1937	1940
Poa pratensis .	56	3	Bouteloua hirsuta .	1.7	3
Sporobolus cryptandrus .	26	39	Carex pennsylvanica .	1	4
Andropogon scoparius .	6	18.5	Koeleria cristata .	1	2.5
Bouteloua curtipendula .	4	14	Cyperus filiculmis .	0	3
Andropogon gerardi .	4	12	Other species* .	0.3	1

* The amount of *Bouteloua gracilis* was based upon the percentage of basal cover. This increased from 5.9 to 16.1 percent.

Based on the number of stems in 1937 and 1940, sand dropseed composed 26 and 39 percent of the vegetation, respectively; little bluestem composed 6 and 18.5 percent; side-oats grama composed 4 and 14 percent; and big bluestem composed 4 and 12 percent. Bluegrass originally composed 56 percent, but finally only 3 percent.

Only 5 of the 23 native forbs that grew in the quadrats were of major importance; together they constituted 90 percent of the total native forb population. Rough pennyroyal (*Hedeoma hispida*), Pursh's plantain (*Plantago purshii*), and daisy fleabane (*Erigeron ramosus*) were the chief forbs in 1937, but all decreased 90 percent or more by 1940 and lost heavily in the unit areas they occupied. Conversely, many-flowered aster and smooth goldenrod increased 480 and 111 percent, respectively, by 1940; they also showed great increase in area. Between 1939 and 1940, they lost considerably in both number of stems and the area occupied.

Only 3 of the 17 species of weeds that grew in the quadrats were of outstanding importance. Horseweed formed 49 percent, peppergrass 44 percent, and little barley 7 percent of the ruderal fraction of the vegetation in 1937. The first decreased 91 percent and the second decreased 99 percent the first year, and both were insignificant thereafter. Little barley (*Hordeum pasillum*) increased nearly 600 percent in 1938, composing 87 percent of the ruderals, and was the chief weed thereafter.

No seedling grasses were found in 1937, but seedlings of grasses (especially sand dropseed) and forbs were abundant in 1939 and 1940.

Relative Production in Pasture and Prairie

The fence on the west side of the enclosure set aside for study was moved—in early spring of 1939—20 feet farther west so as to exclude cattle from a new strip of pasture (Fig. 113). In early spring of 1940, the fence was moved another 20 feet farther west so as to exclude cattle from still another strip of pasture. The amount and the kind of forage that were produced in the pasture that was protected the first year, and in the pasture that was protected a third year, were ascertained in 1939. During 1940, production was measured in pastures that were protected a first, second, and fourth year, and in native prairie.

The marked differences in the development of the tops of the bluestems are shown in Figure 114. Root development of these clumps was not ascertained since the monolith method (to be described later) was not yet devised. However, samples of root systems from similar lots of big bluestem 1 foot wide, 4 feet deep, and 3 inches thick were secured later. Their weights were 9.6, 18.8, and 36.7 grams. This was a decrease of 49 and 76 percent, respectively, from the best-developed clump.

Studies in 1940 on the density and composition of vegetation in plots under the first, second, and fourth year of protection showed

Fig. 113.—A small part of a test pasture (on June 7, 1939). The left portion shows conditions after two years of protection; the right portion, with a great abundance of weeds, had recently been exclosed.

marked differences. Foliage cover averaged 30, 57, and 61 percent under the three conditions, respectively. Grasses and sedges furnished 57, 91, and 97 percent, respectively, of the total vegetation in the several plots. Sand dropseed increased from 49 to 81 percent of the total vegetation, and then decreased to 39 percent. Conversely, little bluestem, big bluestem, and side-oats grama each increased from about 1 percent in the plot that was protected the first year to 18 percent in the plot that was protected a fourth year. Native forbs composed 10, 3, and 1 percent of the vegetation, and ruderals 33, 6, and 2 percent, respectively.

Diameter of the bunches, number of stems per bunch, and the length of stems of little bluestem steadily increased from grazed pasture through pastures that were protected a first, second, and fourth year. Bunches of sand dropseed first increased in size, and then—under greater competition—became smaller in diameter; the stems grew more erect. Otherwise, they behaved in the same fashion as little bluestem.

Yields in 1940 were obtained from adjacent areas of old pasture

Fig. 114.—Bunches of little bluestem (top), selected on June 13, 1940, for average size under first, second, and fourth year of protection. Heights are 3, 5.5, and 9 inches, respectively. Note that the left-most bunch has only a fringe of stems on one side; the second bunch is only one-fourth filled with stems; but the third bunch has a dense stand of stems throughout. Bunches of big bluestem (bottom) selected as in upper figure. Heights are 3.5, 8, and 13 inches. Both photos were taken June 13, 1940.

that were 20 feet wide and 31 rods long and that had been protected from grazing the first, second, and fourth year, and from virgin prairie. Thirty quadrats, located at random under each condition, were clipped at four different intervals. Total yields of prairie grasses in percent—based on the prairie as 100 percent and beginning with the first year of protection—were 8.3, 18.2, 66.7, and 100 percent. Total yields of pasture grasses, based upon the first year of protection as 100 percent, were 100, 125.4, 56.7, and 5.2 percent, respectively. If we state the yields of forbs in prairie as 100 percent, production in the pastures—in order of increasing time of protection—was 61.7, 73.5, and 91.5 percent.

Yields alone do not furnish a proper basis for estimating forage values; hence a utilization factor was determined for each important species. This factor indicated the percentage of the particular species that ordinarily was removed by the livestock in grazing, as determined by rangemen. It averaged 0.77 for the three chief prairie grasses found in pastures; it was 0.50 for sand dropseed but only 0.10 for little barley. The percentage of total yield furnished by each species or group in each pasture was closely estimated, and a utilization of the seasonal yield of prairie and pasture grasses and forbs has been calculated from these data. In tons per acre, the amount of forage that probably would have been utilized by stock from pasture in the several areas protected a first year was 0.49, 0.72, 0.74, and 0.76 tons, respectively. One year of protection more than doubled the amount of the better forage grasses utilized, and three years of protection increased the amount eightfold.

The preceding results were obtained in a four-year period during which severe drought occurred each growing season, but the results apply in general over a vast area of true prairie. They illustrate the manner in which ungrazed pastures return to prairie even during extremely poor years. Under favorable conditions, succession takes place much more rapidly, as the following results will indicate.

Succession from 1941 to 1943

Studies were continued by Weaver and Bruner (1945). Conditions for growth were good in 1941 until the last week in June, when a severe midsummer drought of long duration began. The vegetation became discolored and dry, and growth ceased. Rains revived it late in August, and in 1942 the drought was broken. Rainfall was especially heavy and good water content of soil was maintained at most soil depths throughout the year. The plant cover became much taller and denser. In 1943 there was an excellent summer for growth. Accumulated moisture reserves promoted good yields despite occasional deficit rainfall.

The most marked change in the quadrats dominated by sand dropseed was a steady increase (94 percent) in the number of stems of the dominant until near the end of the drought. This was followed by a gradual decrease as the less xeric grasses regained their vigor.

In quadrats where little bluestem was most abundant, it gained 332 percent by 1940, and even more during the following good years. Sand dropseed gained steadily even after the severe drought of 1939/ 1940, when nearly all other grasses decreased, but thereafter it lost heavily.

Kentucky bluegrass composed 92 percent of the grass in the relict areas that were quadrated in 1937. It suffered a loss of 99 percent during the following dry fall and winter, and had increased only slightly by 1940, but increased very rapidly in 1942, 1943, and 1944. Of the total population of the 43 quadrats, bluegrass comprised 56 percent of the perennial grasses in 1937, but only 3 percent in 1938. Sand dropseed, which constituted 26 percent of the grasses, became more widely distributed than any other species.

Only five native forbs were of major importance. Among these,

FIG. 115.—Views of prairie and pasture in September, 1944. The excellent recovery of the pasture after 3 good years is illustrated .

Aster ericoides and *Solidago missouriensis* both increased greatly, then waned, and later were suppressed.

The foliage cover of vegetation undergoing the first, second, and fourth year of protection was 29, 56, and 61 percent, respectively, in 1940, but this increased to 80 percent or more by 1943.

In 1940, near the end of the drought, sand dropseed still dominated in many places. It grew in mixture with side-oats grama, and over large areas with the bluestems as well. It had been locally replaced by bluestems, which had greatly thickened their stand. The bunches and mats of other prairie grasses also were thicker and more vigorous. Bluegrass had made only slight gains. The patches of goldenrou and aster were somewhat thinned by the grasses, and the plants were dwarfed. The annual weed stage of little barley, Pursh's plantain, and rough pennyroyal had almost disappeared.

After three consecutive good years, a further approach to climax conditions was attained (Fig. 115). Only in a few places was sand dropseed still abundant. The bluestems and side-oats grama dominated, except in parts of lowlands where bluegrass formed a dense sod. There was no bare soil. All the weedy annuals had disappeared; aster and goldenrod also were absent or greatly subdued. An understory of various minor prairie grasses, bluegrass, and forbs was rapidly developing. Under the dense stand of prairie grasses there was a good mulch of fallen debris. However, the absence of certain species of prairie grasses and forbs, which were common in adjacent climax prairie, and the lack of various community relationships indicated that succession was still incomplete.

LATER CHANGES: 1948 to 1953

In 1948 and 1949, the late stage of stabilization in the pasture was studied (Mentzer, 1951).

Large samples of the vegetation of the subsere and of the adjacent contiguous prairie from which it originated were mapped and compared as to composition and structure. A little bluestem type occupied the upper third of the north-facing slope; the big bluestem type had developed on the mid-portion; and a big bluestem–Kentucky bluegrass type was found on the deeper and richer soil of the nearly level lower third of the subsere.

Little bluestem was sparse throughout, but it was more than three times as abundant in prairie as in pasture. Big bluestem in the little bluestem pasture type occurred sparingly and mostly in small bunches. Elsewhere, there was almost as much of this grass in pasture as in prairie. Bluegrass was abundant and widely scattered in pasture, and always present but sparse in prairie. Blue grama occurred only in pasture and side-oats grama was more abundant there. Tall dropseed was abundant in prairie but rare in pasture. Sand dropseed occurred throughout the pasture where it invaded during the drought. It was not found in prairie. Forbs were not abundant in the subsere, even eight years after the drought, despite their large variety and abundance in the prairie.

The average air-dry yield of prairie grasses (1948–49) was 24 percent greater in prairie, but pasture grasses yielded 60 percent less than in the subsere. The yield of forbs was 120 percent greater in prairie.

Although the soil was fully occupied and a good mulch was re-established, the regular cover of climax prairie did not prevail and the understory was poorly developed. A period of many good years will yet be required for dynamic stabilization and the completion of succession.

Study was continued by the present writer until 1953, when the building of a large dam in the area precluded further investigation. The complete story was summarized (Weaver, 1954) as follows.

Plant succession was studied in an old pasture under complete protection at Lincoln, Nebraska, from 1937 to 1953. About half of the vegetation was Kentucky bluegrass. The pasture adjoined a large area of true prairie. Three years of drought had reduced the bluestems to 10 percent of the cover; bluegrass and invading sand dropseed composed 56 and 26 percent, respectively. All but 3 percent of the bluegrass died by the end of the great drought, in 1940; sand dropseed had increased enormously, and—with the xeric side-oats grama—it almost completely dominated. But many less-xeric prairie grasses were scattered throughout and indicated the return of prairie. Big bluestem on lower, more moist slopes, and little bluestem later on upper hillsides, slowly increased, following the drought, over a period of 13 years. Succession was greatly retarded by the presence of the drought population and later by a sod of bluegrass which was rapidly established.

Upper slopes were finally clothed with about 60 percent little bluestem, 30 percent big bluestem, and only 3 percent bluegrass. The lowland now supports extensive pure stands of big bluestem or alternate areas of big bluestem and bluegrass. Forbs at first were largely represented by a few prairie species, which increased greatly, and by numerous annual weeds. Most prairie forbs were exterminated by grazing or by drought. Many have returned only slowly; some are now represented by a few plants; numerous others have not yet reentered the area. This is also true of various grasses. The regenerated prairie is approaching a dynamic equilibrium not greatly unlike the adjacent climax.

This phenomenon of the return of vegetation nearly to its former natural condition—nature's putting back what was there in the first place—is summed up in the term "plant succession." In connection with pastures derived from true prairie, it is of great practical as well as scientific importance, since no pasture at any stage in the degeneration of this prairie continuously produces so great an amount of highly nutritious forage as do the native bluestems.

XIV.

Monolith Method and More Recent Studies

The new method of obtaining samples of root systems is a modification of the direct or trench method that was employed successfully over a period of several decades. By its use, soil is removed from the roots instead of removing the roots from the soil. With the rapid development of soil science and much emphasis on the role of vegetation—especially grasses—in soil formation, a distinct need has arisen for a better understanding of the intimate relation of roots and soil.

THE MONOLITH METHOD

A new method has been devised by which a sample of an entire root system, from soil surface to maximum depth of penetration, may be taken (Weaver and Darland 1949, 1949a). The system can be separated from the soil without injury to the roots and with only slight displacement of individual roots from their natural position. The roots can be examined in the laboratory in relation to the various horizons of the soil profile. The monolith method is of great value in comparing root development and the activities of roots in various soil types and at all soil levels.

A trench about 3 feet wide and 4 to 5 feet long is dug in a site where there is normal development of vegetation. The depth is usually 4 to 6 feet. Beneath the particular sample of grass, which previously was selected and left undisturbed in the side wall, the wall of the trench is made smooth and vertical, as indicated by a plumb line. A long, shallow, wooden box—12 inches wide and 3 inches deep (inside dimensions)—without a top and lacking one end, is employed. It is placed on end, with the closed end downward. The open top is placed against the vertical trench wall, the upper end of the box just reaching the soil surface. An impression of the sides and lower end of the box is made on the wall of the trench by tapping the bottom of the box vigorously with a 4-pound sledgehammer. The box is then removed and the soil column is marked out with butchers' knives that have rigid blades. The soil on the sides and below these marks is removed by means of knives and spades until the monolith protrudes from the trench wall, its sides and bottom projecting outward at least 3 inches. The box is then fitted tightly over the monolith and the bottom and lower end of the box are braced to hold the soil column in place. Finally, the soil on the inner, attached face of the monolith is cut away by working inward with knives and spades from each side. The soil is not cut close to the top of the box, but a V-shaped ridge of soil is left protruding throughout its length. This is a part of the sample when the braces are removed and the monolith is lifted out of the trench. Then the monolith is reduced to exactly 3 inches in thickness by removing the ridge of soil.

The soil is removed from the box by a process of repeated soaking, often for several days, and gentle washing—partly underwater, even when the soil is extremely compact or contains a claypan. A flaring rose nozzle, attached to a garden hose, is employed. During this process one may observe the intimate relations of soil and roots. Roots are unharmed and in their natural position in the water after the soil has been washed away. Efficiency and success are gained only by experience. It is a long process, but one feels well repaid for the time spent when, after 3 to 5 hours, the root system alone is left in almost perfect condition in the bottom of the box. The root system is transferred to a background of black felt and is photographed.

The great detail of root branching that is revealed by this method is illustrated by western wheatgrass. A monolith from a claypan soil where wheatgrass grew in a pure stand was taken on a hillside near Lincoln. The roots had three very different habitats. The A horizon of this soil consisted of a black, well-granulated, silty clay loam that was 12 inches deep. This gave way abruptly to a blocky, prismatic-structure B horizon, which extended to a depth of 28 inches. The clay content increased rapidly with depth, and this hard claypan horizon was removed only with great difficulty in digging the trench. This dark-brown subsoil gave way to yellowish loess of the C horizon, which is a silty clay loam with massive structure and a much smaller clay content. In this yellowish parent material, lime was abundant. The soil was easily removed and the roots were readily separated. The soil broke into blocky pieces and most roots were compressed and flattened between the platy lumps. This condition was maintained to a depth of 4 feet.

Cross sections of the 3 × 12-inch root sample from each soil horizon are shown in Figure 116. There were about 515 roots in the A horizon; 165 roots extended into the B horizon; but the number decreased to 85 at 2 feet in depth. The dry weight of the roots in the B horizon was less than a third of that in the layer above. The C horizon had a root distribution that was distinctly different from that in the A and B horizons. The roots often were flattened on the faces of the small blocks, which cleaved in all directions. Branching therefore occurred in all planes. Hence, after the soil was gently washed away, there was a

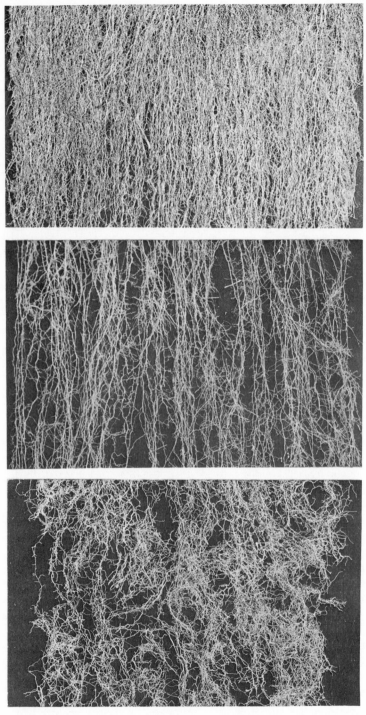

Fig. 116.—Roots of western wheatgrass (*Agropyron smithii*) in the A, B, and C horizons in a claypan soil.

glistening white mass of material with branches that ran outward at all angles. The root habits in the three soil horizons were as different as the environment each presented. Ordinarily, wheatgrass is quite uniform in its root distribution, the amount of roots gradually diminishing with depths of 4 to 6 feet.

Root branching was restricted in the claypan of this Crete soil. Restricted root development in the B horizon was related to severe limitations in the available phosphorous. Such development was not limited in the B horizon of a second Crete soil (half a mile distant) that was characterized by a relatively high level of soluble phosphorous throughout the profile.

A second monolith was taken from a claypan soil (Planosol) near Carleton, 65 miles southwest of Lincoln. The sample was from a small area of Butler silt loam. The surrounding soil type on this nearly level land was Crete silt loam. The monolith was taken in a good stand of western wheatgrass. Recent rains had wet the soil to a depth of about 5 to 6 inches, or nearly through the 7.5-inch A horizon. Exceptions occurred where water entered the large soil cracks and penetrated deeply. The entire B horizon (7.5 to 28 inches) was so hard that the soil was removed only with extreme difficulty. A pick was constantly employed. The roots penetrated the B horizon with difficulty, and probably (except in cracks) only when the soil was moist. Nearly always they were more or less flattened, and in the upper half of the horizon they were more poorly branched than in either the A or C horizon.

The yellowish, less compact, lime-flecked parent material of the C horizon was encountered at a depth of 28 inches. Here, as in the lower third of the B horizon, the branching of the roots was pronounced. Many of the branches were short, flattened, and greatly enlarged. They penetrated the soil in all directions and produced a dense network.

The heavy branching began in the bottom third of the B horizon at 20.5 inches in depth. The weight of the roots above this depth in the B horizon (7.5 to 20.5 inches) actually was less than that in the 13 inches below this level (20.5 to 33.5 inches). The increase in weight at the greater depth was 36.5 percent. Only a few roots penetrated beyond 36 inches.

Associated with the reduced branching below 8 inches was a greatly reduced phosphorus supply and a restricted pore space. The increased branching of roots in the lower B horizon (20 to 28 inches) was associated with an increased percentage of pore space and with a soluble phosphorus content of nearly 2.5 times that present in the region of restricted development (Fox, Weaver, and Lipps, 1953).

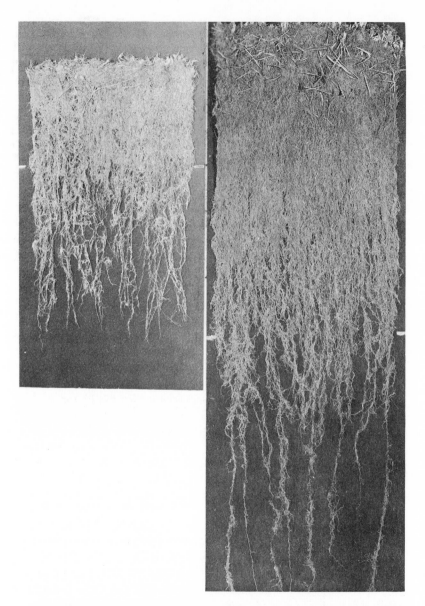

Fig. 117.—Roots of Kentucky bluegrass (*Poa pratensis*) from monoliths taken near Lincoln, in two soil types. The white marks indicate the depth of the A horizon.

This method has many values. It permits bringing root systems in each soil horizon from different soil types together in the laboratory for comparison. Monoliths of Kentucky bluegrass were taken from pure stands of this grass on a nearly level hilltop in Carrington silty clay loam and on a well-drained lowland in Judson silt loam. The root system in the Carrington soil was only 22 inches deep and the bulk of the roots was found in the A horizon, which was only 7 inches thick. Roots in the Judson soil were better developed; some were extremely well branched and reached a depth of 48 inches (see Fig. 117). The deep A horizon (zero to 20 inches) was well filled with a great mass of roots, but here—as in the preceding sample—they became much sparser in the B horizon. The weight of the first root system was less than one-third as great as that of the second. Chemical analyses showed that deep rooting of Kentucky bluegrass occurred in a Judson soil which presented a favorable supply of plant nutrients at all depths in the profile. A Carrington soil, deficient in available phosphorus in the subsoil, produced bluegrass with a shallow root system. Exchangeable potassium and soil nitrogen may also have been limiting factors for root development in this soil (Fox, Weaver, and Lipps, 1953).

Root systems of blue grama at the same stations were very similar to those of bluegrass. The normally deeply rooted big bluestem extended only a few of its roots deeper than 3 feet in the Carrington soil. Its dense root system in the A horizon of the Judson soil weighed twice as much as the entire root system in the Carrington soil. Root depths were 35 and 60 inches.

The root system of blue grama is shown in Figure 118. It was taken near Kearney, Nebraska, in a good stand in a deep soil that is a fairly representative sample of Hastings silt loam, a typical Chernozem. The roots were especially abundant in the deep A horizon (zero to 15 inches). At greater depths, roots were fewer, but they continued to branch profusely to a depth of 4 feet. A few extended beyond the monolith to a depth of 5 feet.

The Colby silt loam, from which a second root system was taken, is present over a large part of the loess hills. Erosion prevents the accumulation of much organic matter and the topsoil is thin and light in color. It ranges from 4 to 8 inches thick. Since the water runs off rapidly (precipitation is about 24 inches), there is no zone of lime accumulation and the surface soil rests directly on the parent material. Even a casual examination of Figure 118 shows that the greatest concentration of roots was in the A horizon (zero to 12 inches) of this shallow solum. Even this layer was relatively poorly occupied: one-fourth less plant material was found here than in the first foot of the preceding sample. Differences at greater depths were even more marked. Depth of penetration was approximately 5.5 feet and half of the root depth occurred below the solum in the parent loess.

In the loess hills of central Nebraska, some low-grade grasslands
have resulted from abandoning cultivated land. A part of the pasture
from which the third sample was taken had been broken and then abandoned for 25 years. As a result of erosion by wind and water, the entire
solum (A and B horizons) had been removed. The remainder was partly
clothed with blue grama, buffalo grass, and a few other grasses.

The monolith was taken on a slope of 11 percent in an area mapped
as Colby silt loam. There was a very thin, poorly developed, new A_1
horizon only 2 to 3 inches deep. The same light-yellow color of soil
prevailed throughout to a depth of 4 feet. The root system is shown in
Figure 118.

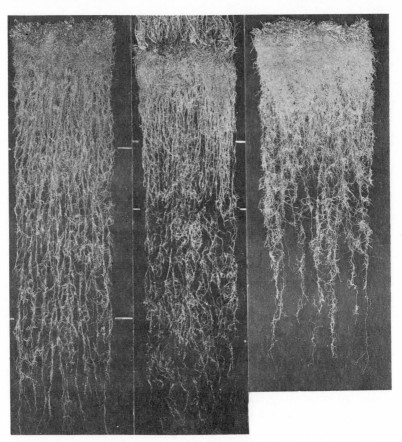

Fig. 118.—Root systems of blue grama (*Bouteloua gracilis*) from Hastings silt loam
(left), from Colby silt loam (center), and from a greatly eroded soil (right). Monoliths
are 4 feet deep, except the last one, which is 3 feet deep. White marks indicate the
depth of the and B horizons.

represents an adaptation for securing water from light showers in the dry, plains soil. Plants of this group were rarely found in true prairie or in the Palouse prairie of the far Northwest.

The third group, the pale-purple coneflower (*Echinacea pallida*) type, consisted of plants with taproots or several main roots which produced relatively few or no branches but penetrated deeply. It composed 20 percent of the forbs.

Forbs with rhizomes, root offshoots, or corms—and with numerous main roots of about equal size—the many-flowered aster (*Aster ericoides*) type, composed the remaining 30 percent.

When a root system is assigned to its type and the depth and lateral spread are stated, a fair conception of its appearance may be had with little further description. The significance of the type in relation to grasses and to different climates has been discussed (Weaver, 1958a).

The Living Network in Prairie Soil

This paper treats of mid-continental soils and the wonderful network of roots within them. Prairie vegetation has clothed these soils for untold centuries. The root systems of more than 20 species of grasses were examined or reexamined in the solum. This network is a permanent feature of prairie. The roots of dominant grasses are not transient but long-lived; the death and decay of individual roots proceeds slowly. Many clumps of sod and bunches of grasses persist below ground for 15 years or more. Moreover, the soil usually contains and acts upon a much more extensive portion of the plant body than does the environment above ground. Prairie vegetation, following or accompanying the physical weathering of parent soil materials, introduced the biological force which was largely responsible for constructional processes in the soil. The accumulating of organic matter, since vegetation returns more to the soil than it takes from it, was a major factor in the formation of both the Brunizem and Chernozem soils. The effects of roots on the parent materials beneath the A and B soil horizons are discussed. While the root system of each species of plant possesses certain inherent characteristics—such as length, depth of penetration, and degree of branching—all of these may be modified by soil conditions. These are topics with which this paper is concerned (Weaver, 1961a).

Typical Western Iowa Prairie

A study of prairies over a wide area began in 1928 and was continued throughout the great drought (1934 to 1941), and intermittently since

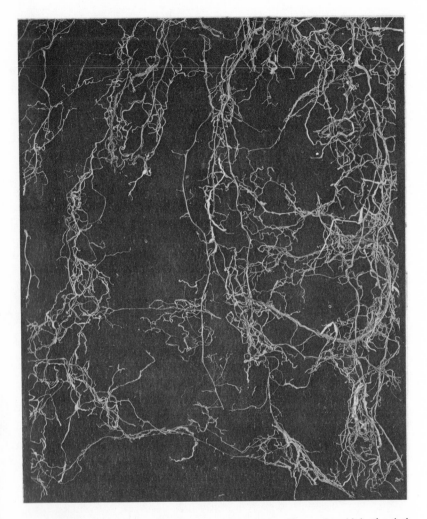

Fig. 119.—Detail of root distribution of brome grass (*Bromas inermis*) in the sixth foot of soil. The width of the sample is 1 foot.

By the use of the monolith method, quantitative data may be obtained on the amount of roots (in grams) in each 6 inches or 1 foot of soil, or in the different soil horizons (see Fig. 119). This also permits expression in percentages of total root weight. The following example, Table 13, is illustrative.

Descriptions have been made of the profiles of 20 soil types, from which 13 species (about 50 monoliths) have been taken. The weight of roots of big bluestem at zero to 6-inch depths ranged from 16.8

TABLE 13

DISTRIBUTION OF ROOT MATERIALS OF BIG BLUESTEM BY
DEPTH, WEIGHT, AND PERCENTAGE

Depth (Inches)	Sharpsburg Silty Clay Loam		Judson Silt Loam	
0–6 . .	33.35 gm.	77.5%	33.69 gm.	68.5%
6–12 . .	4.49	10.4	6.84	13.9
12–24 . .	2.95	6.9	5.15	10.5
24–36 . .	1.35	3.1	2.17	4.4
36–48 . .	0.66	1.5	0.85	1.7
48–60 . .	0.27	0.6	0.51	1.0
Total . .	43.07	100.0	49.21	100.0

DISTRIBUTION OF THE BIG BLUESTEM'S ROOT SYSTEM IN THE
SOIL HORIZONS

Horizon	Sharpsburg Silty Clay Loam			Judson Silt Loam		
A .	12in.	37.84 gm.	87.9 %	20in.	44.17 gm.	89.8 %
B .	48	4.96	11.5	53	4.79	9.7
C .	60	0.27	0.6	60	0.25	0.5

to 53.5 grams in the different soil types (Weaver and Darland, 1949a). The total root weight and distribution of root weight of little bluestem were not greatly different from those of big bluestem, except that the roots of little bluestem were heavier in the surface soil and the root system was 12 to 18 inches shorter. From similar data on other grasses, many comparisons may be derived. The method is very useful in pasture studies (Weaver, 1950).

The monolith method of root study, sometimes with changes in the width and thickness of the sample, has been successfully employed in a study of roots of cultivated crops, such as wheat, corn, sweet clover, etc. This method of root study has also been introduced into Europe (Kmoch, 1957).

Another value of the monolith method, when the width of the sample is increased, is that it reveals the continuity of roots and the relationship of those of one species to those of another. When roots arise from widely spread and abundant rhizomes, as in big bluestem, they pursue an almost vertically downward course. Root distribution is continuous throughout. Even in the deep soil, a remarkably uniform distribution occurs. Where the bunch grass type prevails, as in little bluestem, the widely spreading roots completely occupy the soil. Here roots from other neighboring bunches are included and the mat is again uniform. Only where patches of deep shade are produced by

forbs, as with certain goldenrods and asters, is the root pattern of grasses greatly altered.

Broad monoliths often reveal a type of competition between adjacent grass types that is not evident above ground. Studies by Weaver and Voigt (1950) revealed the underground relationship of big bluestem and Kentucky bluegrass and a similar spreading of roots of switchgrass into the territory occupied by bluegrass. Such invasions are not infrequent at depths of 6 to 24 inches. The roots extend laterally 1 to 2 feet beyond the tops and the rhizomes in the surface soil. The broad monoliths revealed a wide spreading of roots which had not been recorded before in studies of native grass in climax prairie.

The remainder of these studies was made and the following books were written after the writer became Professor Emeritus of Plant Ecology: *North American Prairie* (Weaver, 1954a), *Grasslands of the Great Plains* (Weaver and Albertson, 1956), and *Native Vegetation of Nebraska* (Weaver, 1965).

CLASSIFICATION OF ROOT SYSTEMS OF FORBS

The number of species of forbs in prairie is much greater than that of grasses; about 150 species occur regularly in at least 10 percent of the prairies, and more than 200 others have been listed. A similarly large number occur in the Great Plains. Probably 90 percent or more of the forbs are perennial. Once established, they live—interspersed among the grasses—for many years, and some perhaps live even longer than the grasses. Their underground development in western Iowa, Nebraska, Kansas, and eastern Colorado has been studied by the writer and his students over a period of 40 years (Albertson, 1937; Hopkins, 1951; Tomanek and Alberston, 1957). The root network of forbs supplements that of the grasses and is especially conspicuous in the parent materials. The roots of many forbs attain depths of 15 to 20 or more feet.

Root systems of several mature plants of each of 80 species of forbs were examined and classified. They were of four types. About 25 percent developed taproots with widely spreading branches which originated in the first 3 feet of soil and penetrated deeply with little or no provision for much absorption in the surface soil. This group was designated as the blazing star (*Liatris punctata*) type.

The second group, the broom snakeweed (*Gutierrezia sarothrae*) type, had taproots with maximum spread of abundant laterals in the surface 1 to 2 feet, where extensive branching provided for much absorption. With a subgroup of cacti and several monocots with similar surface-absorbing but fibrous roots, this group also composed 25 percent of the total. There is considerable evidence that this type

that time. This included the grasslands of the western one-third of Iowa, which by 1950 had almost all been plowed or pastured. Indeed, it seemed that all would vanish. Therefore it was decided to write a description of a typical tract of well-watered Iowa prairie that was visited many times in the past 30 years. About 200 acres of prairie that is typical of the rolling to hilly topography of southwestern Iowa were described.

The Brunizem soils were formed from loess on the ridges and gentle slopes, and from glacial till on the steeper slopes. These soils of medium texture absorb most of the 32 or more inches of rainfall and retain it well. They are well aerated and high in both organic matter and mineral nutrients. Thirty to 40 tons of dark, humified organic matter occur per acre in the surface 6 inches of soil. These are our richest prairies. A warm to hot growing season from April to October, with predominantly sunny weather, is followed by cold winters with frozen soil to a depth of 2 or more feet.

The little bluestem community prevailed on the uplands and the big bluestem on lower slopes and lowlands. Vegetation ranged from 5 to 12 inches taller and was more dense than in eastern Nebraska, and yields of hay were one-quarter to one-half ton per acre greater.

A total of about 165 species of forbs were found. They were more abundant and better developed than similar species under lighter rainfall westward. Composition of the vegetation, ranking and classification of forbs, seasonal aspects, prairie in winter, and a general consideration of prairie are topics that are discussed and illustrated (Weaver, 1958*b*).

SUMMARY AND INTERPRETATION OF UNDERGROUND DEVELOPMENT

In further intensive studies of the structure of prairie vegetation, a background of 40 years' experience was employed in defining the root habits of various plant communities on fully developed and stabilized soils. Soil and aerial environment factors that determine root development were used in interpreting community root habits.

Studies of the effects of extreme drought, of the recovery from drought, and of the removal of herbage on root development have aided greatly in the interpretation. They also emphasized the value of a knowledge of the usual community root habit.

The sod-forming tall grasses of lowland communities of true prairie are of the greatest height (5 to 10 feet), have the greatest leaf surface, and produce the largest amount of forage. Their roots are coarsest, and least well branched, but the deepest. They are about as deep (7 to 10 feet) as the stems are tall. Roots do not spread widely just below the soil surface. All of the dominants, except *Elymus canadensis*, are warm-season grasses that grow all summer and flower late.

Upland mid grasses nearly all grow in bunches. They are of intermediate height (2 to 3.5 feet), leaf surface, and amount of forage production. Roots are moderately fine, well branched, and moderately deep. They are about twice as deep (4 to 5.5 feet) as the stems are tall. They are fairly well spread and moderately dense just beneath the soil surface. Grasses from the lowland, when they grow in an upland, are reduced in size and weight. The roots are somewhat finer and more branched, but they penetrate less deeply. There are several cool-season grasses on uplands that flower early, but most are warm-season species that flower late, after a long season for growth.

Grasses of the hardlands of mixed prairie are the smallest in height, leaf surface, and amount of forage production. Mid grasses are usually 1.5 to 2.5 feet high, and short grasses of the understory are 4 to 15 inches high. Both groups are represented by bunch grasses and sodformers. Roots are finest, best branched, and about as deep as those in upland true prairie, but they are 2 to 16 times deeper than the stems are tall. They spread widely and form dense masses in the surface soil, and nearly all are well branched throughout.

The amount of roots, stem bases, and rhizomes in the surface 4 inches of soil is greatest in the lowland of true prairie (3.7 T/A), intermediate in upland true prairie (2.7 T/A), and least in the hardlands in mixed prairie (1.8 T/A). This cover of sod has protected the surface of the earth for centuries against violent physiographic changes and has made possible the formation of soil.

Grasses that grow both in sand and silt loam spread their roots more widely in sand but usually not as close to the soil surface. Several grasses that are found only in sand have excellent rhizomes. The cover of vegetation is more open and the forage production is much less than in upland true prairie, but cover and forage may equal or exceed that of plains hardlands.

VEGETATIONAL CHANGES FROM EAST TO WEST

The western edge of true prairie in northeastern Kansas (the Kansas-Nebraska Drift-Loess Hills) is definitely separated by the Flint Hills region from the western edge of the Kansas-Nebraska Loess Plains (see Fig. 120). In a study of the vegetation, soils, and other environmental factors, the two areas were compared (Weaver, 1960a).

The topography of the two areas is similar except that the hills are more pronounced in the east and more level land alternates with lower hills in the west. Both have highly productive soils. Eastward, on the

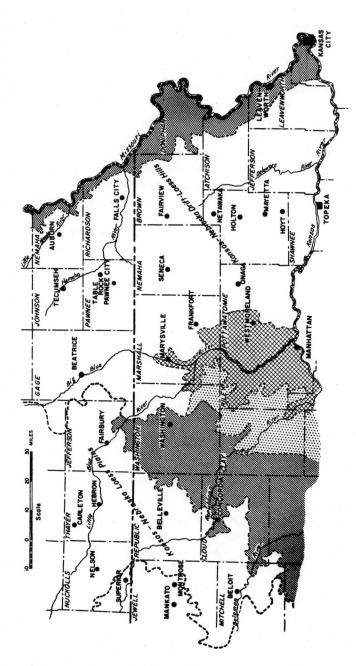

Fig. 120.—Map of the Kansas-Nebraska Drift-Loess Hills and Loess Plains. The Flint Hills and other areas that separate them, and the Loess Hills region along the Missouri River, also are shown. The heavy broken line separates Kansas (southward) from Nebraska.

Brunizems, glacial boulders and limestone outcrops are common. Westward, on the loess plains with Chernozem soils, the bedrock is almost completely covered with loess. Because of reduced precipitation westward—27 or 28 inches compared with 29 to 34 inches in the east— the soils are somewhat less productive. This difference is not due to innate soil fertility. Conditions for growth—such as humidity, wind movement, evaporation, and occurrence and duration of drought—are all less favorable to vegetation in the western area.

In the west, little bluestem occupies a much greater area than in the east, but big bluestem occupies a much smaller area. Foliage cover was not only higher but also much denser eastward. Side-oats grama, June-grass, and various sedges were much more abundant westward. Great Plains grasses, especially blue grama and buffalo grass, which are not observed in prairies eastward, were common. Several other grasses of the Great Plains were represented. Forbs were greatly reduced in number of species and in stature, and numerous species common to the mixed prairie were present.

Extent of Communities and Abundance of Grasses

The extent of communities and the abundance of the most common grasses in the prairie were not a part of the monograph, *The Prairie*, published in 1934. This work was incomplete until 1960. By that time, 63 typical prairies—scattered throughout the 60,000-square-mile area of the central Missouri Valley—had been fully analyzed. Over this region as a whole (see Fig. 29), the little bluestem community occupied only 6 to 11 percent more area in western Iowa than that of big bluestem, but 23 to 58 percent more in the drier areas westward. The presence and amount of *Stipa spartea*, *Koeleria cristata*, and *Sporobolus heterolepis* in various parts of the area were recorded (Weaver, 1960b).

Transition from True Prairie to Mixed Prairie

The middle of the area of transition from true to mixed prairie in South Dakota is about due north at the point where the Missouri River first contacts Nebraska (98° 30' west longitude). It continues due southward through Nebraska and Kansas. Along the Kansas-Nebraska border, this is about 175 miles west of the Missouri River, where annual rainfall approximates 25 inches. Changes in the vegetation, as stated by the writer in *Native Vegetation of Nebraska* (1966), follow.

As one proceeded westward toward this area of transition, the grasses became shorter and more mesic species did not extend as far up

the slopes. Big bluestem and switchgrass were much less widely distributed and showed no marked tendency toward replacing little
bluestem, as had occurred eastward. The sod formed by big bluestem
was much more open and other vegetation was not as completely
excluded by the shade. Yields of hay became smaller as the cover
became more open and the grasses became more dwarfed. Little
bluestem remained dominant over much of the terrain, where one
would ordinarily expect to find big bluestem.

In the area of transition, a zone about 50 miles wide, blue grama
and buffalo grass gradually appeared between the bunches of little
bluestem, and the short grasses became intermingled as an understory
to this grass. Other changes occurred gradually but widely. The density
of short grasses between the bunches of little bluestem and certain
other true prairie grasses greatly increased, and gradually, under
continued grazing, the short-grass pastures were quite in contrast to
the bluegrass pastures eastward. In these dry prairies—whether level
or rolling—switchgrass, nodding wild-rye, and prairie cordgrass found
that suitable habitats were quite limited.

At stations bordering the transition zone on the west, the dominance
of grasses had rather completely changed. Blue grama and buffalo
grass were the most important, and occurred in continuous and sometimes nearly complete stands. Elsewhere they were found as an understory to side-oats grama, western wheatgrass, needle-and-thread,
Junegrass, sand dropseed, and numerous other grasses of mixed prairie.
Big bluestem, Indian grass, and other tall grasses were often found in
deep, moist ravines. Thus the dominant mid grasses were of a more
xeric type, and the prominent understory of short grasses—not found
in true prairie—covered much of the soil without the presence of a
good overstory of mid grasses.

The increased xerophytism from east to west was revealed by the
forbs by their decrease in number of species and smaller stature, and
by the presence of many species of the Great Plains.

The western prairies were not especially conspicuous—as were
those of eastern Nebraska—because of the forb population. This was
partly due to the dwarfness of the plants. Often they attained no more
than half the height of similar species in eastern counties of the state.
Many of the taller and more conspicuous forbs found eastward did not
occur, or were much less frequent. Those that did occur were greatly
reduced in size, and a long list might be given. Conversely, many western
species, which were rarely seen eastward, were present in the area of
transition.

North of the Platte Valley Lowland, the transition from true prairie to mixed prairie is somewhat different because it crosses high hills and intervening valleys. Here species of bluestem prairie have migrated along the valleys to the Sandhills. Short grasses extended eastward along the dry hilltops, while a third community of intermixed species from the eastern prairie and the Great Plains clothed the midslopes (Weaver and Bruner, 1954).

STUDIES IN THE LOESS HILLS

North of the great southward bend of the Platte River in central Nebraska, but southeast of the Sandhills are several thousand square miles of rugged uplands that are known as the loess hills and bluffs. The vegetation is distinctly different from that in sand, and it is very differently distributed than that in the usually more level plains hardlands. In a preliminary survey, the geology, physiography, and climate are discussed. The survey also was concerned with the soil and aerial environment of the abundant, undisturbed native vegetation, and with its composition and behavior in the extensive range lands. It includes a three-year study of grazing types, of grazing patterns throughout the season, and of forage yield and forage consumption in experimental pastures. There are 60 illustrations (Weaver and Bruner, 1948).

A more detailed and complete study of 30 ungrazed prairies in the loess hills was made by Hopkins (1951) during a period of two years, 1948 and 1949. The composition and structure of each of the four types of vegetation—short-grass type, mixed-grass type, mid- and tall-grass type, and the invading wheatgrass type—were studied. Ecotones, plant production, and degeneration under grazing also were considered.

An extensive area of the mixed prairie in the loess hills was studied by Branson and Weaver (1953) to ascertain the degree of degeneration that had resulted from long periods (60 to 70 years) of grazing. It was ascertained that the three types of vegetation demarked different environments or pasture sites. Various degrees of deterioration of the vegetation in each site were revealed in four range-condition classes: excellent, good, fair, and poor. Of 148 ranges examined, only 5 percent were in excellent condition, 17 percent were in good condition, about half were in fair condition, and distinctly poor ranges totaled 28 percent.

Thus the story of plants and their environment in the great midwest presents a wide range of problems.

Bibliography of Literature Cited

Aikman, J. M. 1927. Distribution and structure of the forests of eastern Nebraska. Univ. of Nebraska Studies, 26: 1–75.

Albertson, F. W. 1937. Ecology of mixed prairie in west central Kansas. Ecol. Monog., 7:481–547.

———. 1941. Prairie studies in west central Kansas. Kans. Acad. Science, 41:77–83.

———, G. W. Tomanek, and A. Riegel. 1957. Ecology of drought cycles and grazing intensity on grasslands of Central Great Plains. Ecol. Monog. 27:27–44.

———, and J. E. Weaver. 1942. History of the native vegetation of western Kansas during seven years of continuous drought. Ecol. Monog., 12:23–51.

———, and J. E. Weaver. 1944. Nature and degree of recovery of grassland from the great drought of 1933 to 1940. Ecol. Monog., 14:393–479.

———, and J. E. Weaver. 1945. Injury and death or recovery of trees in prairie climate. Ecol. Monog., 15:393–433.

Aldous, A. E. 1938. Management of Kansas bluestem pastures. Jour. Amer. Soc. Agron., 30:244–253.

Allred, B. W. 1950. Practical grassland management. Sheep and Goat Raiser Magazine, San Angelo, Texas.

Biswell, H. H. 1935. Effects of environment upon the root habits of certain deciduous forest trees. Bot. Gaz., 96:676–708.

———, and J. E. Weaver. 1933. Effect of frequent clipping on the development of roots and tops of grasses in prairie sod. Ecology, 14:368–390.

Blake, A. K. 1935. Viability and germination of seeds and early life history of prairie plants. Ecol. Monog., 5:405–460.

Branson, F. A. 1953. Two new factors affecting resistance of grasses to grazing. Jour. Range Mgt., 6:165–171.

———, and J. E. Weaver. 1953. Quantitative study of degeneration of mixed prairie. Bot. Gaz., 114:397–416.

Carrier, L., and K. S. Bort. 1916. The history of Kentucky bluegrass and white clover in the United States. Jour. Amer. Soc. Agron., 8:256–266.

Clark, O. R. 1937. Interception of rainfall by herbaceous vegetation. Science, 86:591–592.

———. 1940. Interception of rainfall by prairie grasses, weeds, and certain crop plants. Ecol. Monog., 10:243–277.

Clements, F. E., and J. E. Weaver. 1924. Experimental vegetation. Carnegie Inst. Wash. Pub. 355.

———, J. E. Weaver, and H. C. Hanson. 1929. Plant Competition. Carnegie Inst. Wash. Pub. 398, pp. 1–340.

Cratty, R. I. 1929. The immigrant flora of Iowa. Iowa State Col. Jour. Sci., 3:247–269.

Crist, J. W., and J. E. Weaver. 1924. Absorption of nutrients from subsoil in relation to crop yield. Bot. Gaz., 77:121–148.

Darland, R. W., and J. E. Weaver. 1945. Yields and consumption of forage in three pasture types: an ecological analysis. Univ. of Nebraska Conservation and Survey Div. Bul. 27, pp. 1–76.

259

Elder, J. A.   1960.   *In* Soils of the north central region of the United States. North Central Regional Pub. No. 76, Conservation and Survey Div., Univ. of Nebraska, Lincoln.

Flory, E. L.   1936.   Comparison of the environment and some physiological responses of prairie vegetation and cultivated maize. Ecology, 17:67–103.

Fox, R. L., J. E. Weaver, and R. C. Lipps.   1953.   Influence of certain soil-profile characteristics upon the distribution of roots of grasses. Agronomy Jour., 45:583–589.

Garner, W. W., and H. A. Allard.   1923.   Further studies in photoperiodism, the response of the plant to relative length of day and night. Jour. Agr. Res., 23:871–920.

Holch, A. E.   1931.   Development of roots and shoots of certain deciduous tree seedlings in different forest sites. Ecology, 12:259–298.

Hopkins, H. H.   1951.   Ecology of the native vegetation of the loess hills in central Nebraska. Ecol. Monog., 21:125–147.

Howard, A.   1924.   Crop production in India, p. 101, Oxford Univ. Press.

Jacks, G. V.   1944.   The influence of herbage rotations on the soil. *In* Alternate husbandry. Imperial Agric. Bur. Joint Pub. 6.

Jean, F. C., and J. E. Weaver.   1924.   Root behavior and crop yield under irrigation. Carnegie Inst. Wash. Pub. 357.

Kellog, R. S.   1905.   Forest belts of western Kansas and Nebraska. U.S. Dept. of Agriculture, Forest Service Bul. No. 66.

Kmoch, H. G.   1957.   Die Wurzelarbeiten J. E. Weavers und seiner Schule. Zeitschrift fur acker—und pflanzenbau. Band 104, Heft 3 (1957), S. 275–288. Verlag Paul Parey, Berlin und Hamburg.

Kramer, J., and J. E. Weaver.   1936.   Relative efficiency of roots and tops of plants in protecting the soil from erosion. Univ. of Nebraska Conservation and Survey Div. Bul. 12, pp. 1–94.

Mentzer, I. W.   1951.   Studies on plant succession in true prairie. Ecol. Monog., 21:255–267.

Mueller, I. M.   1941.   An experimental study of rhizomes of certain prairie plants. Ecol. Monog., 11:165–188.

———, and J. E. Weaver.   1942.   Relative drought resistance of seedlings of dominant prairie grasses. Ecology, 23:387–398.

Nedrow, W. W.   1937.   Studies on the ecology of roots. Ecology, 18:27–52.

de Peralta, F.   1935.   Some principles of plant competition as illustrated by Sudan grass. Ecol. Monog., 5:356–404.

Pool, R. J., J. E. Weaver, and F. C. Jean.   1918.   Further studies in the ecotone between prairie and woodland. Bot. Surv. of Nebraska, N.S., 2:1–47.

Pound, R., and F. E. Clements.   1900.   The phytogeography of Nebraska (2d ed.) Lincoln, Neb.

Riecken, F. F.   1960.   *In* Soils of the north central region of the United States. North Central Regional Pub. No. 76, Conservation and Survey Div., Univ. of Nebraska, Lincoln.

Robertson, J. H.   1933.   Effect of frequent clipping on the development of certain grass seedlings. Plant Physiol., 8:425–447.

———.   1939.   A quantitative study of true prairie vegetation after three years of extreme drought. Ecol. Monog., 9:433–492.

Shimek, B.   1931.   The relation between the migrant and native flora of the prairie region. Univ. of Iowa, Stud. Nat. Hist., 14:10–16.

Shively, S. B., and J. E. Weaver.   1939.   Amount of underground plant materials in different grassland climates. Univ. of Nebraska Conservation and Survey Div. Bul. 21, pp. 1–68.

Sprague, H. B.   1933.   Root development of perennial grasses and its relation to soil conditions. Soil Sci., 36:189–209.

Steiger, T. L.   1930.   Structure of prairie vegetation. Ecology, 11:170–217.

Stoddart, L. A.   1935.   How long do roots of grasses live? Science, N.S., 81:544.

Stuckey, I. H.   1941.   Seasonal growth of grass roots. Amer. Jour. Bot., 28:486–491.

Thornber, J. J.   1901.   The prairie-grass formation in region I. Rep. Bot. Surv., Univ. of Nebraska, 5:29.

Tolstead, W. L.   1942.   Vegetation of the northern part of Cherry County, Nebraska. Ecol. Monog., 12:255–292.

———.   1947.   Woodlands in northwestern Nebraska. Ecology, 28:180–188.

Tomanek, G. W.   1948.   Pasture types of western Kansas in relation to the intensity of utilization in past years. Kans. Acad. Sci. Trans., 51:171–196.

———, and F. W. Albertson.   1957.   Variations in cover, composition, production, and roots of vegetation on two prairies in western Kansas. Ecol. Monog., 27:267–281.

United States Forest Service.   1937.   Range plant handbook. U.S. Dept. of Agriculture.

Voigt, J. W., and J. E. Weaver.   1951.   Range condition classes of native midwestern pasture: an ecological analysis. Ecol. Monog., 21:39–60.

Weaver, J. E.   1915.   A study of the root systems of prairie plants of southeastern Washington. Plant World, 18:227–248; 273–292.

———.   1917.   A study of the vegetation of southeastern Washington and adjacent Idaho. Univ. of Nebraska Studies, 17:1–133.

———.   1919.   The ecological relations of roots. Carnegie Inst. Wash. Pub. 286, pp. 1–128.

———.   1920.   Root development in the grassland formation. Carnegie Inst. Wash. Pub. 292, pp. 1–151.

———.   1924.   Plant production as a measure of environment. A study in crop ecology. Jour. of Ecology, 12:205–237.

———.   1926.   Root development of field crops. McGraw-Hill, New York, pp. 1–291.

———.   1927.   Some ecological aspects of agriculture in prairie. Ecology, 8:1–17.

———.   1930.   Underground plant development in relation to grazing. Ecology, 11:543–557.

———.   1942.   Competition of western wheatgrass with relict vegetation of prairie. Amer. Jour. Bot., 29:366–372.

———.   1943.   Replacement of true prairie by mixed prairie in eastern Nebraska and Kansas. Ecology, 24:421–434.

———.   1944.   North American Prairie. Amer. Scholar, 13(3):329–339.

———.   1947.   Rate of decomposition of roots and rhizomes of certain range grasses in undisturbed prairie soil. Ecology, 28:221–240.

———.   1950.   Effects of different intensities of grazing on depth and quantity of roots of grasses. Jour. Range Mgt., 3:100–113.

———.   1950a.   Stabilization of midwestern grassland. Ecol. Monog., 20:251–270.

———.   1954.   A seventeen-year study of plant succession in prairie. Amer. Jour. Botany, 41:31–38.

———.   1954a.   North American prairie. Johnsen Pub. Co., Lincoln, Neb., pp. 1–348.

———.   1958.   Summary and interpretation of underground development in natural grassland communities. Ecol. Monog., 28:55–78.

———.   1958a.   Classification of root systems of forbs of grassland and a consideration of their significance. Ecology, 39:393–401.

———.   1958b.   Native grasslands of southwestern Iowa. Ecology, 39:733–750.

———.   1960.   Flood plain vegetation of the central Missouri Valley and contacts of woodland with prairie. Ecol. Monog., 30:37–64.

———.   1960a.   Comparison of vegetation of the Kansas-Nebraska drift-loess hills and loess plains. Ecology, 41:73–88.

Weaver, J. E. 1960*b*. Extent of communities and abundance of the most common grasses in prairie. Bot. Gaz. 122:25–33.

————. 1961. Return of midwestern grassland to its former composition and stabilization. Occas. Papers, Adams Center Ecolog. Studies, 3:1–15.

————. 1961*a*. The living network in prairie soils. Bot. Gaz., 123:16–28.

————. 1963. The wonderful prairie sod. Jour. Range Mgt., 16:165–171.

————. 1965. Native vegetation of Nebraska. Univ. of Nebraska Press, Lincoln, pp. 1–185.

————, and F. W. Albertson. 1936. Effects of the great drought on the prairies of Iowa, Nebraska, and Kansas. Ecology, 17:567–639.

————, and F. W. Albertson. 1939. Major changes in grassland as a result of continued drought. Bot. Gaz., 100:576–591.

————, and F. W. Albertson. 1940. Deterioration of grassland from stability to denudation with decrease in soil moisture. Bot. Gaz., 101:598–624.

————, and F. W. Albertson. 1940*a*. Deterioration of midwestern ranges. Ecology, 21:216–236.

————, and F. W. Albertson. 1943. Resurvey of grasses, forbs, and underground plant parts at the end of the great drought. Ecol. Monog., 13:63–117.

————, and F. W. Albertson. 1956. Grasslands of the Great Plains; their nature and use. Johnsen Pub. Co., Lincoln, Neb., pp. 1–395.

————, and W. E. Bruner. 1927. Root development of vegetable crops. McGraw-Hill, New York, pp. 1–351.

————, and W. E. Bruner. 1945. A seven-year quantitative study of succession in grassland. Ecol. Monog., 15:297–319.

————, and W. E. Bruner. 1948. Prairies and pastures of the dissected loess plains of central Nebraska. Ecol. Monog., 18:507–549.

————, and W. E. Bruner. 1954. Nature and place of transition from true prairie to mixed prairie. Ecology, 35:117–126.

————, and F. E. Clements. 1938. Plant ecology (2d ed.). McGraw-Hill, New York, pp. 1–601.

————, and R. W. Darland. 1947. A method of measuring vigor of range grasses. Ecology, 28:146–162.

————, and R. W. Darland. 1948. Changes in vegetation and production of forage resulting from grazing lowland prairie. Ecology, 29:1–29.

————, and R. W. Darland. 1949. Quantitative study of root systems in different soil types. Science, 110(No. 2850):164–165.

————, and R. W. Darland. 1949*a*. Soil-root relationships of certain native grasses in various soil types. Ecol. Monog., 19:303–338.

————, and T. J. Fitzpatrick. 1932. Ecology and relative importance of the dominants of tall-grass prairie. Bot. Gaz., 93:113–150.

————, and T. J. Fitzpatrick. 1934. The prairie. Ecol. Monog., 4:109–295.

————, and E. L. Flory. 1934. Stability of climax prairie and some environmental changes resulting from breaking. Ecology, 15:333–347.

————, and W. W. Hansen. 1939. Increase of *Sporobolus cryptandrus* in pastures of eastern Nebraska. Ecology, 20:374–381.

————, and W. W. Hansen. 1941. Native midwestern pastures, their origin, composition and degeneration. Univ. of Nebraska Conservation and Survey Div. Bul. 22, pp. 1–93.

————, and W. W. Hansen. 1941*a*. Regeneration of native midwestern pastures under protection. Univ. of Nebraska Conservation and Survey Div. Bul. 23, pp. 1–91.

————, H. C. Hanson, and J. M. Aikman. 1925. Transect method of studying woodland vegetation along streams. Bot. Gaz., 80:168–187.

————, and G. W. Harmon. 1935. Quantity of living plant materials in prairie soils in relation to runoff and soil erosion. Univ. of Nebraska Conservation and Survey Div. Bul. 8, pp. 1—53.

————, and W. J. Himmel. 1929. Relation between the development of root system and shoot under long- and short-day illumination. Plant Physiol., 4:435–457.

————, and W. J. Himmel. 1930. Relation of increased water content and decreased aeration to root development in hydrophytes. Plant Physiol., 5:69–92.

————, and W. J. Himmel. 1931. The environment of the prairie. Univ. of Nebraska Conservation and Survey Div. Bul. 5, pp. 1–150.

————, V. H. Hougen, and M. D. Weldon. 1935. Relation of root distribution to organic matter in prairie soils. Bot. Gaz., 96:389–420.

————, F. C. Jean, and J. W. Crist. 1922. Development of activities of roots of crop plants. Carnegie Inst. Wash. Pub. 316, pp. 1–117 (14 plates).

————, and J. Kramer. 1932. Root system of *Quercus macrocarpa* in relation to the invasion of prairie. Bot. Gaz., 94:51–85.

————, J. Kramer, and M. Reed. 1924. Development of root and shoot of winter wheat under field environment. Ecology, 5:26–50.

————, and I. M. Mueller. 1942. Relative drought resistance of seedlings of dominant prairie grasses. Ecology, 23:387–398.

————, and W. C. Noll. 1935. Comparison of runoff and erosion in prairie, pasture, and cultivated land. Univ. of Nebraska Conservation and Survey Div. Bul. 11, pp. 1–37.

————, and N. W. Rowland. 1952. Effects of excessive mulch on development, yield, and structure of native grassland. Bot. Gaz., 114:1–19.

————, L. A. Stoddart, and W. Noll. 1935. Response of the prairie to the great drought of 1934. Ecology, 16:612–629.

————, and A. F. Thiel. 1917. Ecological studies in the tension zone between prairie and woodland. Bot. Surv. of Nebraska, N.S., 1:1–60.

————, and G. W. Tomanek. 1951. Ecological studies in a midwestern range: the vegetation and effects of cattle on its composition and distribution. Univ. of Nebraska Conservation and Survey Div. Bul. 11, pp. 1–82.

————, and J. Voigt. 1950. Monolith method of root-sampling in studies on succession and degeneration. Bot. Gaz., 111:286–299.

————, and E. Zink. 1945. Extent and longevity of the seminal roots of certain grasses. Plant Physiol., 20:359–379.

————, and E. Zink. 1946. Annual increase of underground materials in three range grasses. Ecology, 27:115–127.

————, and E. Zink. 1946a. Length of life of roots of ten species of perennial range and pasture grasses. Plant Physiol., 21:201–217.

Weinmann, H. 1948. Investigations on the underground reserves of South African grasses. So. African Sci., 2:12–15.

Yocum, W. W. 1937. Root development of young Delicious apple trees as affected by soils and by cultural treatments. Res. Bul. 95, Agr. Expt. Sta., Lincoln, Neb., pp. 1–55.

The following doctoral theses in plant ecology at the University of Nebraska (1924–1952) discuss ecological conditions that range from Canada through Montana and Nebraska to Oklahoma, Texas, and Arizona, as well as in Puerto Rico.

Bruner, W. E. 1931. The vegetation of Oklahoma. Ecol. Monog., 1:99–188.

Cottle, H. J. 1931. Studies in the vegetation of southwestern Texas. Ecology, 12:105–155.

Coupland, R. T. 1950. Ecology of mixed prairie in Canada. Ecol. Mong., 20:271–315.

Dyksterhuis, E. J.   1946.   The vegetation of the Fort Worth prairie. Ecol. Monog., 16:1–29.

Fredricksen, M. T.   1938.   Comparison of the environment and certain physiological activities of alfalfa and prairie vegetation. Amer, Midl. Nat., 20:641–681.

Garcia-Molinari, O.   1952.   Grasslands and grasses of Puerto Rico. Bul. 102, Univ. of Puerto Rico Agr. Expt. Sta., Rio Piedras, P.R.

Hanson, H. C.   1924.   A study of the vegetation of northeastern Arizona. Univ. of Nebraska Studies, 24:85–178.

Heady, F. H.   1950.   Studies on bluebunch wheatgrass in Montana and height-weight relationships of certain range grasses. Ecol. Monog., 20:55–81.

Marsh, I. M.   1940.   Water content and osmotic pressure of certain prairie plants in relation to environment. Univ. of Nebraska Studies, 40:1–44.

Noll, W. C.   1939.   Environment and physiological activities of winter wheat and prairie during extreme drought. Ecology, 20:479–506.

Norris, E. L.   1939.   Ecological study of the weed population of eastern Nebraska. Univ. of Nebraska Studies, 39:29–91.

Pavlychenko, T. K.   1942.   Root systems of certain forage crops in relation to the management of agricultural soils. National Research Council (Canada), No. 1088, Ottawa.

Stoddart, L. A.   1935.   Osmotic pressure and water content of prairie plants. Plant Physiol., 10:661–680.

Index

A horizon, of Brunizems, 3; of Chernozems, 5; true prairie, 27
Acer negundo (boxelder), 122, 127, 134–135, 139; competition, 114–116; in drought, 140–141
Acer saccharinum, 114–116
Achillea lanulosa, 155
Achillea millefolium (yarrow), 202
Actinomycetes, 3
Aesculus glabra, 127
Agropyron cristatum (crested wheatgrass), 89
Agropyron smithii (western wheatgrass), 16, 17, 244, 245, 257; decay of, 95; depths of water absorption, 92; in drought: 150, 151, 153, recovery, 178–181, 183, 185, 187, 188, 192, 194, resistance, 187; grazing reaction, 199; length of root life, 89; range percentage composition, 217
Agrostis hyemalis, 199
Aikman, J. M., 110, 121, 124, 127, 133
Air temperatures in true prairie, 28
Albertson, Dr. F. W., 29, 140, 149, 151, 154, 156, 160, 165, 171, 173–176, 186–188, 194, 251
Aldous, A. E., 198
Alfalfa, erosion and, 74; irrigation effect, 97–98; rainfall interception, 68; runoff and, 75, 76
Allard, H. A., 96
Allionia linearis, 19
Allred, B. W., 221
Amaranthus retroflexus (redroot), 107
Amaranthus spp, 188
Ambrosia elatior (annual ragweed), 203
Ambrosia psilostachya, 203; in drought, 155
Ambrosia trifida (giant ragweed), 107
Amelanchier canadensis, 108
American elder, 108, 122, 127
American elm, 122, 127, 134–135; competition, 114–116; in drought, 141, 144
Amorpha canescens (lead plant), 14, 15,

59–60, 215; distribution of, 59–62; in drought, 172; grazing reaction, 202; importance of, 59–62
Amorpha fragrans, 108
Amorpha fruticosa (indigo bush), 122, 127
Amorpha nana, 62
Andropogon furcatus, 167, 187
Adropogon gerardi (big bluestem), 6, 8, 9, 11, 32–37, 233, 236, 250, 257; cattle preference for, 220; competition, 105, 107, 112, 114; decay of, 94–95; depths of water absorption, 92; in drought: 148–149, 151–153, 162–165, 167–168, 171–173, recovery, 178–184, 186, 191, 192, 194; erosion and, 71; grazing reaction, 197–198; length of root life, 89; palatability of, 197–198; rainfall interception, 68, 69; range percentage composition, 217; rhizomes of, 77, 78; root increase (annual), 86–88; seminal root system of, 85, 86; top increase (annual), 86–88; top removal development, 225; vegetation percentages, 55, 56, 57
Andropogon hallii (sand bluestem), 21, 22
Andropogon scoparius (little bluestem), 6–10, 12, 21, 22, 32–39, 64, 65, 79, 233, 236, 238, 257; cattle preference for, 220; competition, 105–106, 112, 113; decay of, 94–95; in drought: 148, 151–154, 162–165, 167–169, recovery, 178–184, 186, 191, 193, 194, resistance, 187; grazing reaction, 197–199; length of root life, 89; palatability of, 198–199; rainfall interception, 68; range percentage composition, 217; root increase (annual), 86–88; top increase (annual), 86–88; top removal development, 225; vegetation percentages, 55, 56, 57
Anemone canadensis (Canada anemone), 61
Anemone caroliniana, 193
Annual ragweed, 203
Anogra cinerea, 23

265